Astronomers' Univ

For other titles published in this series, go to
http://www.springer.com/series/6960

John Wilkinson

New Eyes on the Sun

A Guide to Satellite Images and Amateur Observation

Dr. John Wilkinson
4 Ross Drive
Castlemaine, Victoria, 3450
Australia
johnwbmw@gmail.com

ISSN 1614-659X
ISBN 978-3-642-22838-4 e-ISBN 978-3-642-22839-1
DOI 10.1007/978-3-642-22839-1
Springer Heidelberg Dordrecht London New York

Library of Congress Control Number: 2011939837

Cover figure: The Solar Probe Plus spacecraft with its solar panels folded into the shadows of its approach to the Sun. Credit: NASA/JHU/APL.

Springer is part of Springer Science+Business Media (www.springer.com)

Preface

The Sun is the only star in the universe that is close enough for us to see its features in intricate detail. Astronomers have observed the Sun over many centuries from Earth but in recent decades they have gained a new understanding via scientific instruments in satellites (space probes). The remarkable dynamical phenomena occurring on the Sun, such as eruptions of matter in the form of prominences, filaments, spicules, coronal outbursts, and gigantic flares make the study of the Sun extremely fascinating and interesting.

Activity on the Sun is the most energetic in our Solar System. Without the Sun's energy in the form of heat and light, life on Earth could not exist. The Sun also affects our weather, climate and communication systems, and an understanding of it is essential to daily life.

In the past, observing the Sun has been left to academics with specialised instruments, since solar observation has been unsafe because of the risk of eye damage. It is now possible for amateur astronomers to safely observe the various solar phenomena using special hydrogen-alpha telescopes that are not too expensive. Amateurs can now make a positive contribution to science by monitoring the Sun as professionals do.

Most of the new information about the Sun has come from recent satellites that observe the Sun on a daily basis using visible, x-ray and ultraviolet wavelengths of electromagnetic radiation. Amateurs can also access the solar images taken by satellites via the internet and smart phones. This book helps readers interpret and understand what these images are showing about the Sun. Readers will enjoy comparing their own solar telescope observations with those produced by space probes such as SDO, SOHO, Hinode and STEREO.

The main purpose of this book is to present some of the fascinating solar phenomena, in their full glory to readers through a variety of illustrations, photographs and easy to understand text.

Solar astronomy is becoming very popular among amateur astronomers as well as academics. Amateurs can use special telescopes to observe the Sun every day. Many surface features show changes in appearance every hour – this is different to night time observing where the appearance of objects remains fairly static night after night. This book also provides the latest space probe information for people interested in studying the Sun and more importantly, it bridges the gap between advanced astrophysics of the Sun and elementary knowledge about the Sun.

In recent years a new term, space weather, has come into vogue to describe the effect the Sun's radiation has on Earth and the environment of space near Earth. The study of space weather is critical to our survival and to an understanding of our environment.

Linked to space weather is the effect of the Sun on Earth's climate. Many scientists think that we are undergoing climate change and believe that the Sun may be a cause of this change.

The final chapter of this book looks at the Sun as a star. There are many different types of stars each with particular characteristics. Stars also go through a life cycle whereby they grow, change and die. Readers will enjoy learning about the evolution and fate of the Sun as a star.

Hopefully you will find enjoyment in this book and improve your understanding of the Sun and enjoy the growing hobby of Solar Astronomy.

Dr John Wilkinson

Acknowledgments

The author and publisher are grateful to the following for the use of photographs in this publication: National Aeronautics and Space Administration (NASA), European Space Agency (ESA), Goddard Space Flight centre (GSFC), Johnson Space Centre (JSC), Japan Aerospace Exploration Agency (JAXA), the Royal Swedish Academy of Science and LMSAL. Many of the photographs have been taken from recent space probes used to study the Sun up close, such as SOHO, STEREO, TRACE and SDO. Credit for each photograph is provided with the captions. Where credit is not given, the photo is the author's.

The author would like to acknowledge the assistance of Stephen Ramsden (USA) for his advice and help with some of the photographs of solar observing equipment.

While every care has been taken to trace and acknowledge copyright, the author apologises in advance for any accidental infringement where copyright has proven untraceable. He will be pleased to come to a suitable arrangement with the rightful owner in each case.

Note: The websites used in this book were correct at the time of publication.

Contents

1. Warming to the Sun

The Sun is the only star in the universe that astronomers on Earth can study in detail. There is no other star close enough to show features of the order of a few hundred kilometres. In the past decade or so, astronomers have modified their understanding of the Sun's processes because of new discoveries made by solar satellites. This chapter introduces the reader to the key features and behaviours of the Sun in preparation for an in-depth study of the Sun.

The Solar System

The Sun, planets and their moons form a family of bodies called the Solar System. We know that the Sun is at the centre of the Solar System and that the major planets orbit the Sun in nearly circular orbits. Our understanding of the Solar System has changed dramatically over the centuries as bigger and better telescopes were developed and new data of planetary motions was collected. Although people knew the Sun and planets of the solar system existed, little was known about the nature of these worlds until space probes containing scientific instruments were sent into space to explore these objects from close range.

The Sun is a star because it emits its own heat and light through the process of thermonuclear fusion. Planets differ to stars in that they do not emit much heat and they do not have enough mass for fusion to occur in their core. We see the planets because they reflect sunlight. The Sun and planets are spherical in shape, and are held in that shape by a force called gravity.

The planets orbit the Sun in much the same plane, and because of this they all appear to move across the sky through a

J. Wilkinson, *New Eyes on the Sun*, Astronomers' Universe,
DOI 10.1007/978-3-642-22839-1_1, © Springer-Verlag Berlin Heidelberg 2012

Fig. 1.1 The Sun as seen from the space shuttle as it orbits around Earth.

common narrow path. Observed from a position above the Sun's north pole, all the planets orbit the Sun in an anti-clockwise direction (Fig. 1.1).

The time taken by a planet to orbit the Sun is called its **period of revolution**. A planet's period depends on its distance from the Sun: the further a planet is from the Sun, the slower its speed and the longer its period.

The four "inner planets" (Mercury, Venus, Earth and Mars) are smaller, denser and rockier than the "outer planets" of Jupiter, Saturn, Uranus, and Neptune. The inner planets are also warmer and rotate more slowly than the outer planets. The outer planets are gaseous planets, containing mostly hydrogen and helium with some methane and ammonia. They are also cold and icy with deep atmospheres.

During the 1800s, astronomers discovered a large number of small, rocky bodies orbiting the Sun between Mars and Jupiter.

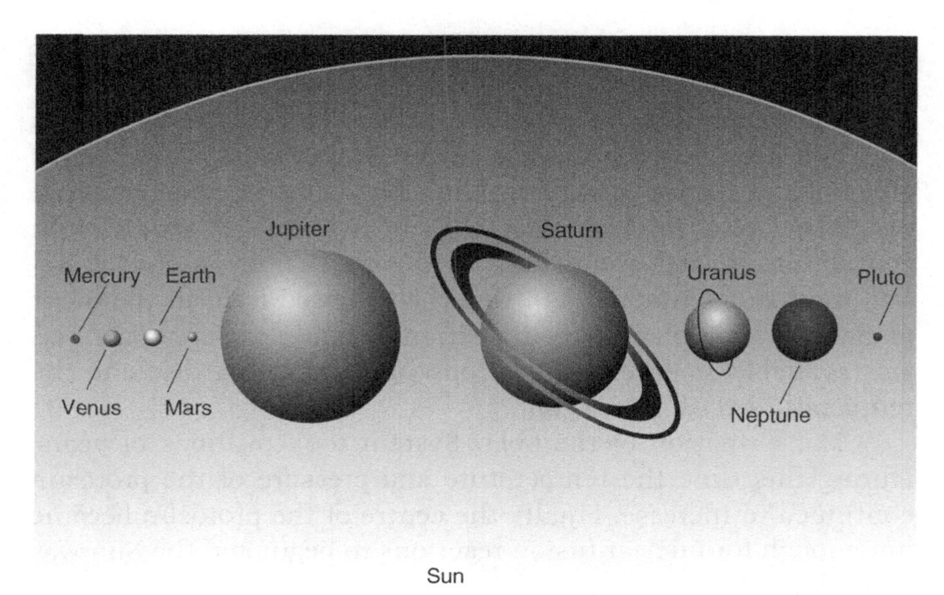

Fig. 1.2 The planets in front of a disc of the Sun, with all bodies drawn to the same scale (Fig. 1.2).

Bodies such as Ceres, Pallas and Vesta, that had been thought of as small planets for almost half a century, were classified as **asteroids**.

Formation of the Solar System

The Solar System is thought to have formed about 4.5 billion years ago from a vast cloud of very hot gas and dust called the **solar nebula**. This cloud of interstellar material began to condense under its own gravitational forces. As a result, density and pressure at the centre of the nebula began to increase, producing a dense core of matter called the **protosun**. Collisions between the particles in the core caused the temperature to rise deep inside the protosun.

The planets and other bodies in the Solar System formed because the solar nebula was rotating. Without rotation, everything in the nebula would have collapsed into the protosun. The rotating material formed a flat disc with a warm centre and cool edges. This explains why nearly all the planets now rotate in much the same plane. Astronomers have found similar discs around other stars.

As the temperature inside the protosun increased, light gases like hydrogen and helium were forced outward while heavy elements remained closer to the core. The heavier material condensed to form the inner planets (which are mainly rock containing silicates and metals), while the lighter, gaseous material (methane, ammonia and water) condensed to form the outer planets. Thus a planet's composition depends on what material was available at different locations in the rotating disc and the temperature at each location.

The formation of the Solar System took millions of years. During this time the temperature and pressure of the protosun continued to increase. Finally the centre of the protosun became hot enough for nuclear fusion reactions to begin and the Sun was born (Fig. 1.3).

Stars like our Sun can take 100 million years to form from a nebula. Radioactive data of the oldest material in our Solar System suggests it is about 4.6 billion years old.

Much of the debris leftover from the formation of the Solar System is in orbit around the Sun in two regions – the Asteroid Belt and the Kuiper Belt. The Asteroid belt lies between Mars and Jupiter while the Kuiper belt is a region beyond Neptune.

The Asteroid Belt contains over a million rocky bodies. About 2,000 of these have very elliptical orbits, some of which cross Earth's orbit. Stony fragments ejected during asteroid collisions are called meteoroids. Some of these meteoroids enter the Earth's atmosphere and burn up releasing light; such bodies are called meteors. Large meteors that impact with the ground form craters like those seen on the surface of the Moon.

The Kuiper Belt is a bit like the Asteroid Belt, except that it is much farther from the Sun and it contains thousands of very cold bodies made of ice and rock. Objects in this outer region take up to 200 years to orbit the Sun.

There are other smaller objects on the outer edge of Solar System, in a region called the Oort cloud. This roughly spherical cloud also contains many objects left over from the formation of the Solar system. Comets are icy bodies that originate from these outer regions of the Solar System. Many comets have highly

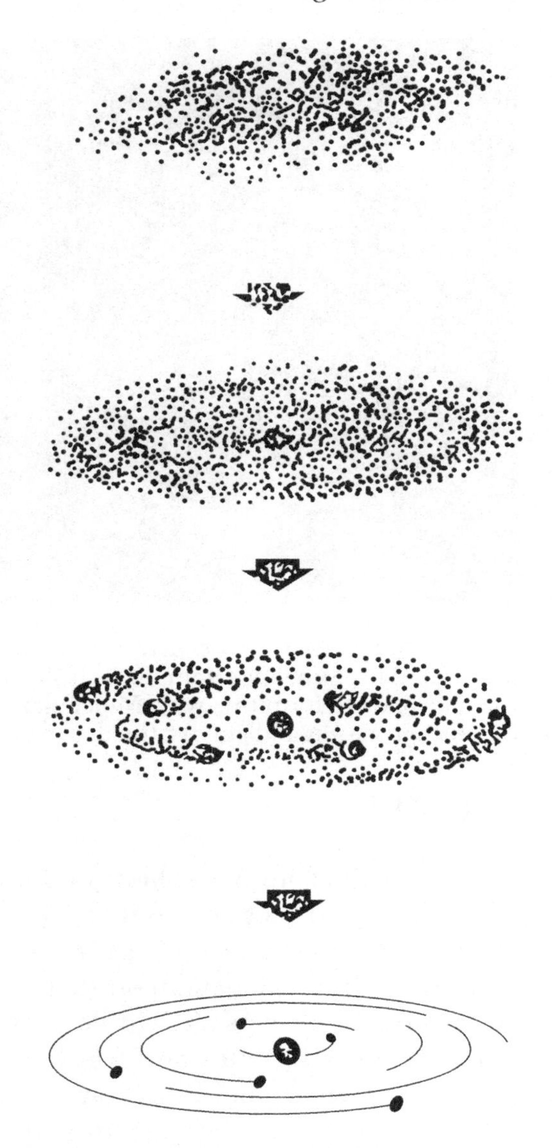

Fig. 1.3 Stages in the formation of the Solar System: (**a**) A slowly rotating cloud of interstellar gas and dust begins to condense under its own gravity. (**b**) A central core begins to form a protosun. A flattened disc of gas and dust surrounds the protosun, and begins to rotate and flatten. (**c**) The planets begin to condense out of the flattened disc as it rotates. (**d**) The planets have cleared their orbit of debris.

elongated orbits that occasionally bring them close to the Sun. When this happens the Sun's radiation vaporizes some of comet's icy material, and a long tail is seen extending from the comet's head. Each time they pass the Sun, comets lose about 1% of their mass. Thus comets do not last forever. Comets eventually break apart, and their fragments often give rise to many of the meteor showers we see from Earth (Fig. 1.4).

Fig. 1.4 Star formation occurring in a cloud of interstellar gas and dust (nebula). The Sun, for example, is thought to have taken about 30–50 million years to form this way. Molecular clouds provide the raw material for star formation and gravity the driving force.

The Sun

The Sun is the dominant object in the solar system because it is by far the largest object and contains 99% of the system's mass. It is positioned at the centre of the solar system and its gravitational pull holds all the planets in orbit. The Sun is an average sized star about 4.6 billion years old. Unlike planets, stars produce their own light and heat by burning fuels like hydrogen and helium in a process known as nuclear fusion. Stars have a limited life and the Sun is no exception – it is about half way through its life cycle of about ten billion years.

The Sun is one of over 100 billion stars that make up the Milky Way galaxy. The Milky Way galaxy is spiral in shape, and the Sun is positioned about halfway from the centre. The Milky Way is about 100,000 light years in diameter and 15,000 light years thick. You can see parts of the Milky Way as a band of cloud that stretches across the night sky. Within the Milky Way, the Sun is

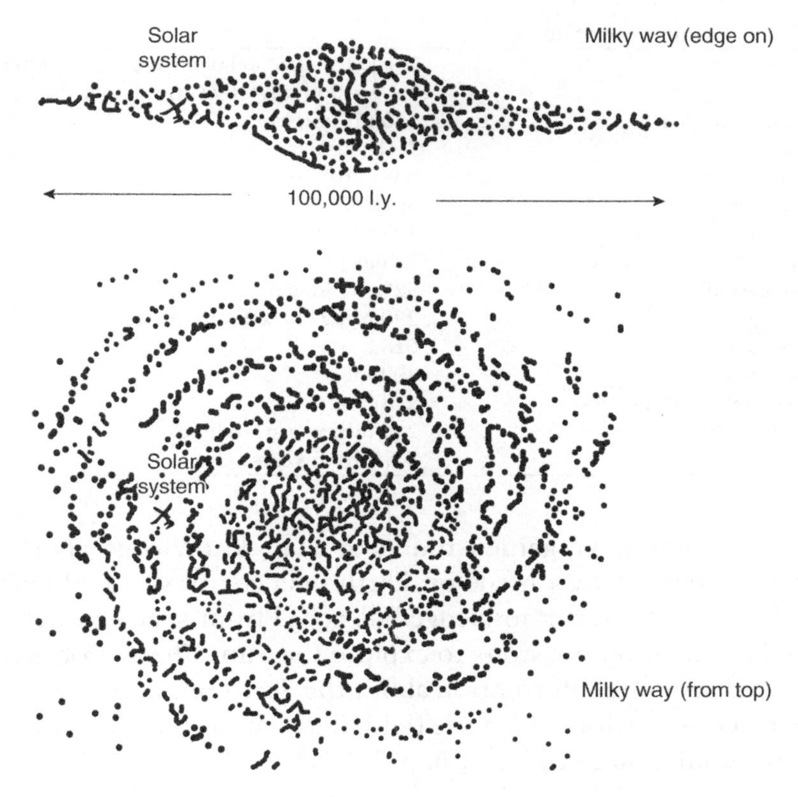

Fig. 1.5 Position of the Sun in the Milky Way galaxy.

moving at 210 km every second, and takes 225 million years to complete one revolution of the galaxy's central mass of stars (Fig. 1.5).

Did You Know?
It takes about 8 min and 20 s for light to travel from the Sun to Earth. This means that sunlight falling on the ground now, left the Sun 8 min 20 s ago. Why is this?

The speed of light is 300,000 km/s. The Sun is 150 million kilometres away from Earth, so it takes light 150,000,000 divided by 300,000 = 500 s or 8 min 20 s to reach Earth.

How long does it take light to reach Earth from the next nearest star?

Astronomers use the light years (ly) as a measure of distance in the universe. One light year is the distance that light travels in one Earth year. In one year light travels a distance of 9.5 million kilometres. The nearest star to our solar system (apart from the Sun) is Proxima Centauri at about 4.2 light years. In a scale model with the Sun and Earth 30 cm apart, Proxima Centauri would be 82 km away. Because Proxima Centauri is 4.2 ly away from Earth, it takes 4.2 years for light from that star to reach us. Thus when you look up into the night sky at this star, the light you see left the star 4.2 years ago. You are really looking into the past.

Table 1.1 Details of the Sun

Mass	2.0 × 10^{30} kg (300,000 times Earth's mass)
Diameter	1.4 million km (109 times larger than Earth)
Distance from Earth	150 million km
Density	1,410 kg/m^3
Luminosity	3.9 × 10^{26} J/s
Surface temperature	5,600°C
Interior temperature	15 million °C
Equatorial rotation period	25 days
Composition	92% hydrogen, 7.8% helium
Surface gravity	290 N/kg (29 × Earth gravity)
Escape velocity	618 km/s
Photosphere thickness	400 km
Chromosphere thickness	2,500 km
Core pressure	250 billion atmospheres
Age	4.6 billion years

Scientists have gained much of their knowledge about the Sun from observation made on Earth over many years. However, much of our current knowledge has come from space probes that have been sent on missions to explore the Sun. These probes have provided accurate information about the Sun's temperature, atmosphere, composition, magnetic field, flares, prominences, sunspots and internal dynamics (see Chap. 2) (Table 1.1).

Composition of the Sun

The Sun is a huge ball of plasma – hot ionised gas, and contains over 300,000 times more mass than the Earth. The Sun's diameter of 1.4 million km far exceeds Earth's diameter of only 12,760 km. Even the biggest planet – Jupiter, is only one tenth the diameter of the Sun.

The main elements present in the Sun are hydrogen (92%), followed by helium (7.8%), and less than 1% of heavier elements like oxygen, carbon, nitrogen and neon. Table 1.2 shows the composition of the Sun constructed from analysis of the solar spectrum. The analysis comes from the lower layers of the Sun's atmosphere but it is thought to be representative of the entire Sun with the exception of the Sun's core. About 67 elements have been detected in the solar spectrum.

Table 1.2 Composition of the Sun

Element	Abundance (percentage of total number of atoms)	Abundance (percentage of total mass)
Hydrogen	91.2	71.0
Helium	8.7	27.1
Oxygen	0.078	0.097
Carbon	0.043	0.40
Nitrogen	0.0088	0.096
Silicon	0.0045	0.099
Magnesium	0.0038	0.076
Neon	0.0035	0.058
Iron	0.0030	0.14
Sulfur	0.0015	0.040

The Sun is believed to be entirely gaseous with an average density 1.4 times that of water. Because the pressure in the core is much greater than at the surface, the core density is eight times that of gold, and the pressure is 250 billion times that on Earth's surface.

Almost all the Sun's mass is confined to a volume extending only 60% of the distance from the Sun's centre to its surface.

Source of Energy and Luminosity

The Sun produces a hundred million times more energy than all the planets combined. Just over half this energy is in the form of visible light, with the rest being infrared (heat) radiation. Only about a billionth of the Sun's energy reaches us here on Earth.

According to the gaseous model of the Sun, the Sun's energy comes from the "burning" of its hydrogen via the process of nuclear fusion. In this process hydrogen (H) nuclei react together to make helium (He) nuclei. This process can take a number of alternate steps but the end result is the same. In the following steps the numbers after each element's symbol indicate the number of particles (protons plus neutrons) in the element's nucleus.

Step 1: H-1 + H-1 $\rightarrow$ H-2 + positron + neutrino
Step 2: H-2 + H-1 $\rightarrow$ He-3 + gamma ray photon
Step 3: He-3 + He-3 $\rightarrow$ He-4 + 2H-1

In the first step, two hydrogen-1 nuclei combine to make a hydrogen-2 nuclei with the emission of a positron and neutrino. The neutrino escapes from the Sun directly, but the positron collides with an electron and they annihilate each other, with a release of energy. In the second step, a hydrogen-2 nuclei fuses with a hydrogen-1 nuclei within about 10 s to form a helium-3 nuclei, with the emission of a gamma ray. In the third step, two helium-3 nuclei fuse to make one nucleus of ordinary helium-4 plus two hydrogen-1 nuclei.

During the above reactions, some mass is lost and it is this mass that is converted into energy (according to Einstein's $E = mc^2$ equation). Every second the Sun converts over 600 million tonnes of hydrogen into helium. At present the central hydrogen is depleting at a rate of five million tonnes every second. In spite of this enormous depletion of fuel from the Sun's core, the Sun should last for another 4.5 billion years.

The energy from fusion raises the temperature and pressure in the core. Because ions and electrons in the core are moving at very high speeds and are densely packed together, they don't travel far before striking other particles. In each collision the particles exert forces on each other. They then rebound and collide with other particles. The result of these frequent collisions is an outward force sufficient to counterbalance gravity. The balance between the inward force of gravity and the outward force from the gas pressure is called hydrostatic equilibrium.

The luminosity of a star is an indication of the total amount of energy it produces every second. This rate depends on the core temperature and pressure of the star, which in turn depends on its mass. The Sun's luminosity is 3.9×10^{26} J/s. Throughout its life the Sun has increased its luminosity by about 40% and it will continue to increase for some time.

Interior of the Sun

The Sun's interior is a giant nuclear reactor generating the vast amounts of energy needed to support the star against the inwards pull of gravity. Like other stars of similar mass and temperature, the Sun's interior contains three zones:

- The core,
- The radiative zone, and
- The convective zone.

The Core

The core is the central region of the Sun where nuclear reactions convert hydrogen into helium. These reactions release the energy that ultimately leaves the Sun as visible light. For these reactions to take place a very high temperature is needed. The temperature close to the centre is about 15 million degrees Celsius and the density is about 160 g/cm^3 (i.e. 160 times that of water). Both the temperature and density decrease outwards from the centre of the Sun. The core occupies the innermost 25% of the Sun's radius. At about 175,000 km from the centre the temperature is only half its central value and the density drops to 20 g/cm^3.

The Radiative Zone

Surrounding the core is the radiative zone. This zone occupies 45% of the solar radius and is the region where energy, in the form of gamma ray photons, is transported outward by the flow of radiation generated in the core. The high-energy gamma ray photons are knocked about continually as they pass through the radiative zone, some are absorbed, some re-emitted and some are returned to the core. It may take the photons a 100,000 years to find their way through the radiative zone. At the outermost boundary of the radiative zone, the temperature is about 1.5 million degrees, and the density is about 0.2 g/cm^3. This boundary is called the interface layer or tachocline. It is believed that the Sun's magnetic field is generated by the magnetic dynamo in this layer. The changes in fluid flow velocities across the layer can stretch magnetic field lines of force and make them stronger. There also appears to be sudden changes in chemical composition across this layer.

The Convective Zone

The outermost zone is called the convective zone, because energy is carried to the surface by a process of convection. It extends from a depth of about 210,000 km up to the visible surface and occupies about 30% of the Sun's radius. In this zone, plasma gas, heated by the radiative zone beneath, rises in giant convection currents to the surface, spreading out, cooling, and then shrinking – similar to the boiling of water in a pot. Rising cells of gas are visible on the surface as a granular pattern. The granules are around 1,000 km in diameter. The convection cells release energy into the Sun's atmosphere. At the surface the temperature is around 5,600° and density is practically zero.

Once the plasma gas reaches the surface of the Sun, it cools and settles back into the Sun to the base of the convection zone, where it receives more heat from the top of the radiative zone. The process then repeats itself. The photons escaping from the Sun, have lost energy on their way up from the core and changed their wavelength so most emission is in the visible region of the electromagnetic spectrum.

The lower temperatures in the convective zone allow heavier ions, (such as carbon, nitrogen, oxygen, calcium, and iron), to hold onto some of their electrons. This makes the material more opaque so that it is harder for radiation to get through. This traps heat that ultimately makes the fluid unstable and it starts to "boil" or convect.

Helioseismology and the Interior

Astronomers have learnt more about the Sun's internal structure by studying sound-like waves oscillating in its interior. The process, called helioseismology, is similar to geologists using earthquake waves to study the Earth's interior. The Sun vibrates at a variety of frequencies, similar to a ringing bell. These sound vibrations were first noticed in 1960, and can be detected with instruments that measure the Doppler shifts of spectral lines belonging to the gases. Results show the Sun's surface moves in and out by about 10 km every 5 min. Slower vibrations with periods ranging from 20 min to nearly an hour were discovered in

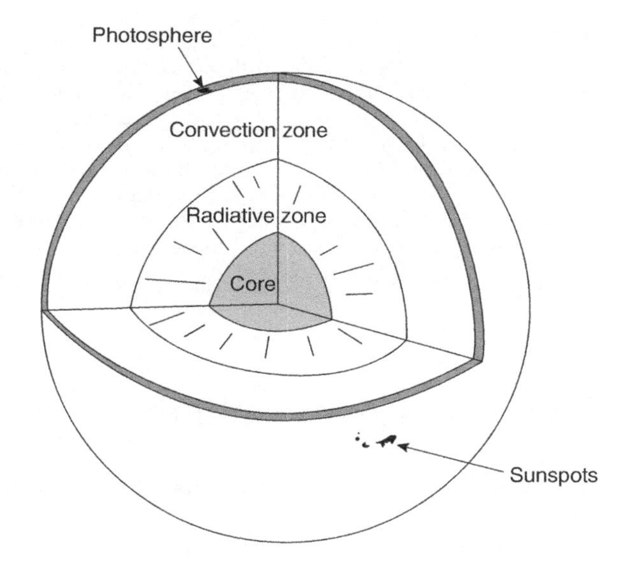

Fig. 1.6 Interior structure of the Sun. Energy is transferred by radiation in the inner regions, and by convection in the outer region.

the 1970s. Temperature, composition, and motions deep in the Sun influence the oscillation periods and yield insights into conditions in the solar interior. One result of these studies is that the convective zone is twice as thick as early solar models predicted. Another finding is that below the convective zone, the Sun apparently rotates like a rigid body (Fig. 1.6).

Did You Know?

The neutrinos produced in the Sun's core are bundles of energy, but they are different in character from photons, which are perceived as electromagnetic energy. Neutrinos have amazingly high penetrating power because they practically do not interact with matter once they are created and they are very difficult to detect. In the Sun 2×10^{38} neutrinos are produced every second, and these come streaming out of the Sun unhindered into space. One would expect millions of neutrinos would hit each square centimetre of Earth every second. Detection of these neutrinos would provide scientists with valuable information about the Sun's interior. Raymond Davis of the Brookhaven National Laboratory was the first to devise a large neutrino detector in the early 1960s. The devise consisted of a huge tank filled with perchloroethylene cleaning fluid buried deep in a mine in South Dakota. The mine shielded the fluid from cosmic rays, since cosmic rays can produce neutrinos in the Earth's atmosphere. Most solar neutrinos would pass through Davis' tank with no effect. On rare occasions, though, a neutrino strikes the nucleus of the chlorine atoms in the cleaning fluid and converts one of its neutrons into a proton, creating a radioactive atom of argon. The rate at which argon is produced is correlated with the number of neutrinos from the Sun arriving at the Earth. On average, solar neutrinos create one radioactive argon atom every 3 days in Davis' tank. This was about one third of the expected result. So where are the missing solar neutrinos? Some astronomers suggested that there are other types of neutrinos that we cannot detect. It is now believed that many solar neutrinos change form en-route from the Sun (to muons or taus) and as a result can't be detected once they reach Earth. The first direct evidence of this came in 2001 from the Sudbury Neutrino Observatory in Canada. Scientists at the observatory were able to detect all types of neutrinos coming from the Sun, and could distinguish between electron neutrinos and the other two forms. After extensive statistical analysis, it was found that about 35% of the arriving solar neutrinos are electron neutrinos, with the others being muon- or tau neutrinos. The total number of detected neutrinos agrees quite well with the earlier predictions from nuclear physics, based on the fusion reactions inside the Sun.

The Sun's Atmosphere

The Sun's atmosphere contains three regions that vary in temperature and density (Fig. 1.7).

- The photosphere
- The chromosphere
- The corona

The Photosphere

The photosphere is the lowest of the three layers comprising the Sun's atmosphere. Because the upper two layers are transparent to most wavelengths of visible light, we see through them down to the photosphere. We cannot, see through the shimmering gases of

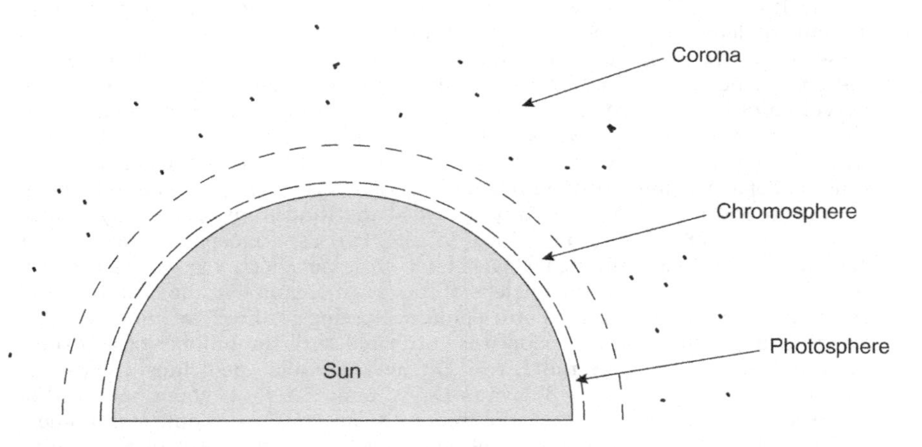

Fig. 1.7 Structure of the Sun's atmosphere (not to scale).

the photosphere, and so everything below the photosphere is regarded as the Sun's interior.

The photosphere (meaning "sphere of light") is a thin shell of hot, ionised gases or plasma about 400 km thick, the bottom of which forms the visible surface of the Sun. Most of the energy radiated by the Sun passes through this layer. From Earth, the surface looks smooth, but it is actually turbulent and granular because of convection currents. Material boiled off from the surface of the Sun is carried outward by the solar wind.

The density of the photosphere is low by Earth standards, about 0.1% as thick as the air we breathe, and its average temperature is only 5,600°C. The composition of the photosphere is, by mass, 74.9% hydrogen and 23.8% helium. All heavier elements account for the less than 2% of the mass.

High-resolution photographs of the Sun's surface reveal a blotchy pattern called granulation. Granules are around 1,000 km in diameter, are convection cells in the Sun's photosphere. By measuring the wavelengths of spectral lines in various parts of the granules, astronomers have found that hot gas rises upward in the centre of a granule. As it cools, the gas radiates its energy, in the form of visible and electromagnetic radiation, out into space. The cooled gas then spills over the edges of the granules and plunges back into the Sun along the boundaries between granules. A granule's centre is brighter than its edges because the centre is at a higher temperature (see photo in Fig. 1.8).

The photosphere appears darker around the limb of the Sun, than it does toward the centre of the solar disc. This phenomenon is called limb darkening. This effect is due to the spherical nature of the Sun and the different depths to which we look into the photosphere. Because we are looking at a spherical body, our angle of observation changes when we move from the limb towards the centre. Also, when we look at the centre of the solar disc, we look at the bottom of the photosphere where it is hotter (6,500°C) and brighter. When we look at the edges we see light from higher in the photosphere where it is slightly cooler (5,600°C) and less bright. The light reaching us from the limb also passes through a more heavily absorbing thickness of atmosphere and is accordingly less bright than that emanating from the centre of the disc. See Fig. 1.9.

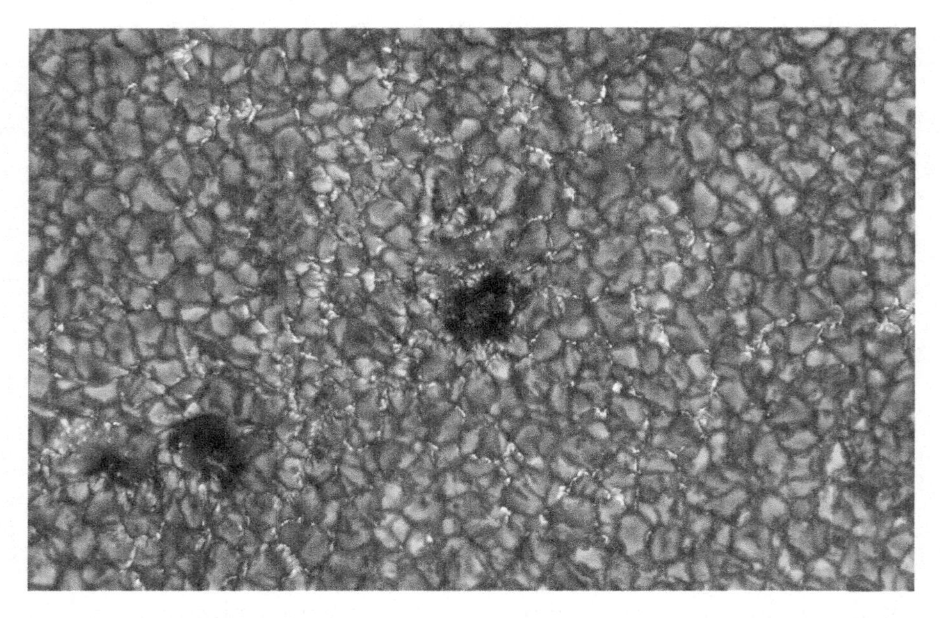

Fig. 1.8 Solar granulation is the result of hot gas rising from the interior in giant convection cells and then spreading out and dispersing.

A number of features can be observed in the photosphere with a solar telescope (contains a special filter to reduce the intensity of sunlight to safely observable levels). These features include the sunspots and faculae. These features will be introduced next and covered in more detail in Chap. 3.

The surface of the Sun contains dark areas called sunspots. Sunspots appear dark because they are cooler than the surrounding photosphere – about 3,500°C compared to 5,600°C. They radiate only about one fifth as much energy as the photosphere.

Sunspots vary in size from 1,000 km to over 40,000 km and they move slowly across the surface of the Sun as the Sun rotates. Their lifetime seems to depend on their size, with small spots lasting only several hours, and larger spots persisting for weeks or months. The rate of movement of sunspots can be used to estimate the rotational period of the Sun. At the equator, sunspots take about 25 days to move once around the Sun. At the poles sunspots take about 36 days to go around the Sun. Sometimes sunspots appear in isolation, but often they are seen in groups.

Sunspots are directly linked to the Sun's intense magnetic fields. Such spots are areas where concentrated magnetic fields

Fig. 1.9 Sunspots on the surface of the Sun on 11th March 2011. Large sunspots contain a *dark centre* (umbra) with a *grey surrounding area* (penumbra). The darker the region, the lower the temperature. Notice also the effect of limb darkening (Credit: NASA/SDO).

break through the hot gases of the photosphere. These magnetic fields are so strong that convective motion beneath the spots is greatly reduced. This in turn reduces the amount of heat brought to the surface as compared to the surrounding area, so the spot becomes cooler. Data obtained from space probes have shown that the strength of the magnetic fields around sunspots is thousands of times stronger than the Earth's magnetic field. Sometimes these magnetic fields change rapidly releasing huge amounts of energy as solar flares into the Sun's atmosphere.

The number of sunspots visible on the Sun tends to vary from a minimum to a maximum and back again in a solar cycle of about 11 years (although this may vary from 7 to 17 years in any particular cycle). During the period of each sunspot cycle, the Sun's magnetic field reverses, so after 22 years the magnetic field returns to its original orientation. See Chap. 3 for more details on sunspots.

Faculae are another visible phenomenon of the photosphere associated with the Sun's magnetic field. Faculae are irregular patches or streaks brighter than the surrounding surface. They are clouds of incandescent gas in the upper regions of the photosphere. Such clouds often precede the appearance of sunspots. The photo in Fig. 3.14 shows some faculae as bright patches against the surrounding surface. See Chap. 3 for further details on faculae.

The Chromosphere

The layer of the Sun's atmosphere immediately above the photosphere is the chromosphere (sphere of colour). This layer of tenuous gas has a density much less than that of the photosphere. It is about 2,500 km thick with a temperature that varies from 6,000°C just above the photosphere to about 20–30,000°C at its top. The chromosphere is more visually transparent than the photosphere. It appears a reddish-pink because of its emission is mainly gaseous hydrogen alpha light. This colour may be seen during a total solar eclipse, when the chromosphere is seen briefly as a flash of colour just as the visible edge of the photosphere disappears behind the Moon. Astronomers can also observe the chromosphere through special narrow wavelength filters tuned to the H-alpha spectral line (see Chap. 6).

Observations of the flash spectrum during a total solar eclipse indicate the main elements present in the chromosphere are hydrogen (most abundant), helium, and smaller amounts of sodium, magnesium, calcium and iron.

For reasons not fully understood, the temperature of the chromosphere is higher than that of the photosphere. Some scientists think that acoustic wave turbulence is the source of this higher temperature in the lower and middle regions of the chromosphere, while magneto-hydro-dynamic waves contribute to heating the upper chromosphere.

Spicules are one of the most common features within the chromosphere. They are long thin jets of luminous gas, projecting upwards from the photosphere. Spicules rise to the top of the chromosphere at about 72,000 km/h and then sink back down again over the course of about 10 min (see Fig. 1.10).

The chromosphere also contains fibrils which are dark protrusions or horizontal wisps of gas similar to spicules but with about twice the duration. Another feature is the prominences, which are gigantic plumes of gas rising up through the chromosphere from the photosphere. They are suspended above the photosphere by magnetic fields and can reach heights of up to 150,000 km. Prominences come in a variety of shapes and sizes; some only last for a few hours while others persist for days. These flame-like structures can be seen projecting out from the limb of the Sun during total solar eclipses. They can also be seen through H-alpha solar telescopes. Temperatures in the prominences can reach 50,000°C.

Fig. 1.10 Numerous spicules can be seen in this photograph of the Sun's chromosphere. The spicules are jets of gas that surge upwards into warmer regions of the Sun's outer atmosphere (Credit: NASA/SDO).

The chromosphere is also the site of solar flares. Flares sometimes appear as a sudden brightening of an existing plage (a bright patchy region within the chromosphere). The energy output of a flare can be enormous. They are also known to eject particles of matter in addition to electromagnetic radiation. Flares are associated with sunspot groups. During a solar flare, temperatures in a compact region sour to five million degrees. Such flares usually last for 20 min. Ultraviolet and x-ray radiation from a flare take about 8 min to reach Earth, while high energy particles from them arrive a day or two later. These particles interfere with radio communications and often produce intense auroras in the Earth's atmosphere.

The chromosphere is also visible in the light emitted by ionised calcium, in the violet part of solar spectrum at a wavelength of 3934 Å (the Calcium K-line). Special solar filters are available to view the Sun in this wavelength.

Extending further out from the chromosphere, the solar atmosphere becomes particularly tenuous. There is a thin layer between the chromosphere and corona called the transition zone. The layer contains mostly ionised hydrogen and heavier elements, which emit a lot of ultraviolet light.

Beyond this zone the temperature of the atmosphere begins to increase markedly to the hot corona, which forms the outer part of the atmosphere.

The Corona

The corona (meaning "crown") is the upper layer of the Sun's atmosphere, and it extends several million kilometres from the top of the chromosphere into space around the Sun. There is no well-defined upper boundary to the corona.

The corona can only be seen during a total solar eclipse or through a special telescope called a coronagraph, when the photosphere is blocked out. The corona appears as a pale white glowing area around the Sun. Its shape varies with the strength of the sunspot cycle.

The corona appears more extended around the solar disc when there are few sunspots on the Sun. When there are a lot of sunspots, the corona is restricted to the equatorial or sunspot zones. This restriction is connected with the increase in magnetic activity during periods of many sunspots.

Temperatures in the corona reach as high as one million degrees Celsius because of interactions between gases and the photosphere's strong magnetic fields. A spectrum of the corona contains emission lines of a number of ionised elements in the plasma state. For example, one prominent line is that of ionised iron.

Although the temperature is extremely high, the density of the corona is very low, about ten trillion times less dense than the air on Earth at sea level. This low density partly accounts for the dimness of the corona. The Sun's strong gravity prevents most of the ionised gas in the corona from escaping into outer space. Even so, the Sun ejects around a million tonnes of matter each second, mostly as charged particles in the solar wind. See Fig. 1.11.

The solar wind is an erratic flow of highly ionised gas particles that are ejected into space from the Sun's upper atmosphere. This wind has large effects on the tails of comets and even has

Fig. 1.11 White light picture of the solar corona as seen during a total solar eclipse. The corona is the white crown surrounding the Sun (Credit: NASA/UCAR/NCAR High Altitude Observatory).

measurable effects on the trajectories of spacecraft. Near the Earth, the particles in the solar wind move at speeds of about 400 km/s. These particles often get trapped in Earth's magnetic field, especially around the poles, and produce geomagnetic storms called auroras.

The corona is optically thin at visible wavelengths. Thus visible light passes through the corona and any coronal features cannot be seen against the solar disc. To see coronal features, one has to make use of soft x-rays or extreme ultraviolet (EUV) radiations. Some of these features will be briefly discussed below and in greater detail in Chap. 3.

The corona emits a lot of x-ray radiation. X-ray pictures reveal a very blotchy, irregular inner corona, with numerous x-ray bright spots that are hotter than the rest of the corona. Temperatures in these bright spots occasionally reach four million degrees. These occur over the surface of the Sun in active regions, although the number in equatorial zones seems higher. X-ray bright points change with time by spreading out, and decreasing in brightness.

EUV radiation is used to observe coronal holes. A coronal hole is a large region in the corona that is less dense and cooler than its surroundings. Such holes may appear at any time during a solar cycle but they are most common during the declining phase of the cycle. Coronal holes are the source of open magnetic field lines that move way out into space. They allow denser and faster "gusts" of the solar wind to escape the Sun at speeds around 800 km per second. They are sources of many disturbances in Earth's ionosphere and geomagnetic field.

On x-ray and EUV images of the Sun, coronal holes appear as dark regions because they do not emit much radiation at these wavelengths, while the surrounding regions do (see Fig. 1.12).

A coronal mass ejection or CME is an expulsion of a part of the corona and ionised particles into space. Such events can represent the loss of several billion tonnes of matter from the Sun at speeds between 10 and 1000 km/s. Some CME's are triggered by solar flares and are associated with strong magnetic fields in the corona. Sometimes, clouds of ejected particles are carried by the solar wind towards Earth. The particles get trapped in the Earth's magnetic field and interfere with satellite motion, communications and power systems (see Fig. 1.13 and Chap. 4).

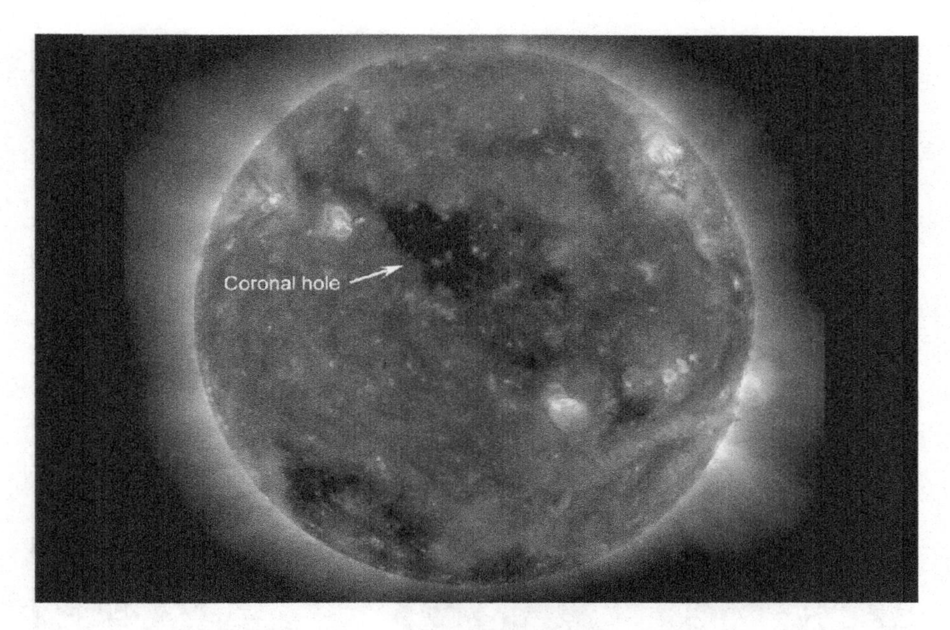

Fig. 1.12 A coronal hole as seen in ultraviolet light by the SDO space probe on 10th January 2011. Such holes are the source of open magnetic field lines that move way out into space. They are also source regions of the fast solar wind, which is characterised by a relatively steady speed of about 800 km/s (Credit: NASA/SDO/AIA).

Other Solar Radiation

The Sun gives off many kinds of radiation besides visible light and heat. These radiations include radio waves, ultraviolet rays and x-rays. Space probes that orbit the Sun make observations and take pictures in the different wavelengths of electromagnetic radiation.

The Sun's chromosphere and corona are also emitters of radio waves. These were first recorded in 1942 during World War II by British radars as "radio noise". Such radio emissions often originate from sunspots and produce what we call "solar storms". These radio waves can be collected via radio telescopes on Earth. Observations of the Sun using radio waves provide information different to that obtained from visible wavelengths because the propagation of the two types of radiations are different. Coronal gas is transparent to visible light but is opaque to radio waves.

Ultraviolet rays are electromagnetic waves with a shorter wavelength than visible light. They are invisible to the human

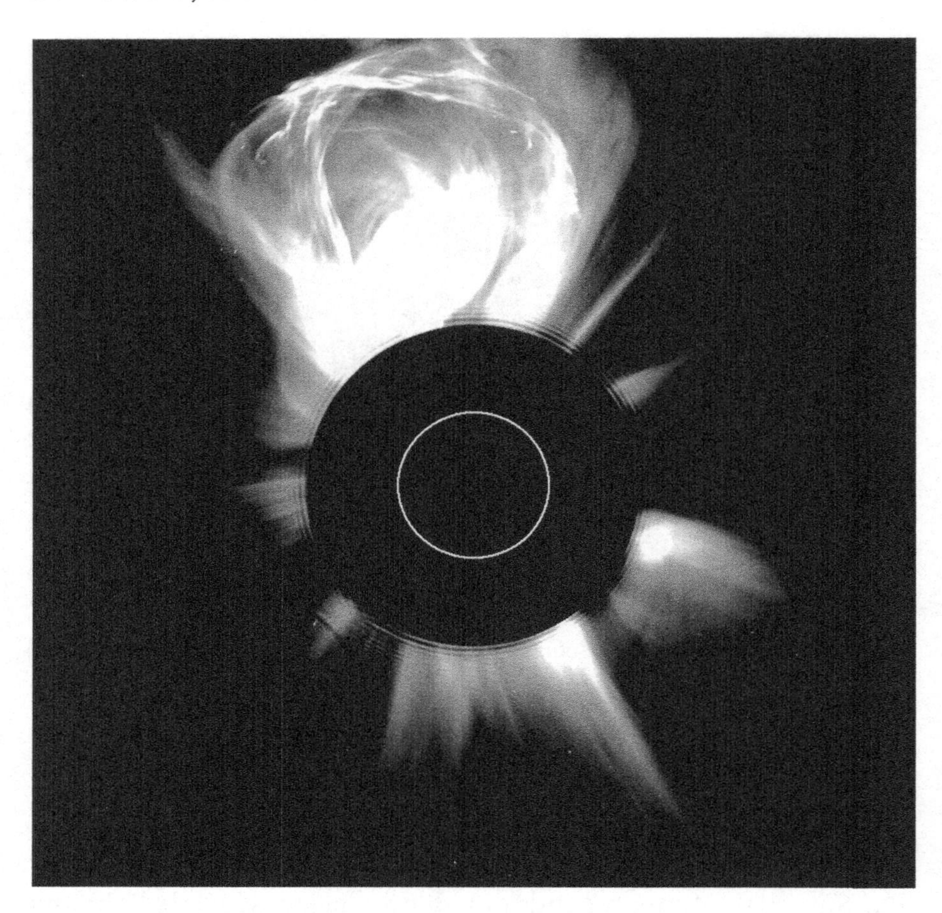

Fig. 1.13 In March 2000 an erupting filament lifted off the active solar surface and blasted this enormous bubble of magnetic plasma into space (a coronal mass ejection). Direct light from the sun is blocked in this picture of the event with the sun's relative position and size indicated by a *white circle* at *centre*. The field of view extends two million kilometres or more from the solar surface (Credit: NASA/ESA/SOHO).

eye. The Sun gives off more ultraviolet radiation during times of increased solar activity. The Earth's atmosphere absorbs much of this radiation. Scientists have divided the ultraviolet part of the spectrum into three regions: the near ultraviolet, the far ultraviolet, and the extreme ultraviolet. The three regions are distinguished by how energetic the ultraviolet radiation is, and by the "wavelength" of the ultraviolet light, which is related to energy. The near ultraviolet, abbreviated NUV, is closest to optical or

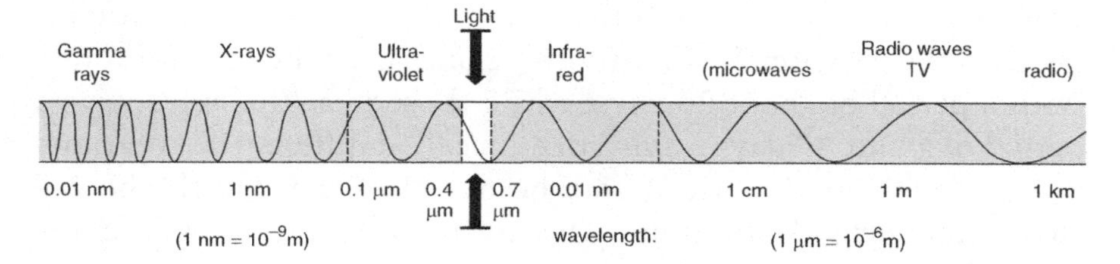

Fig. 1.14 The Sun can be observed using different types of electromagnetic radiation each with their own range of wavelengths. Short wavelength radiation (more energetic) is shown on the *left*, while long wavelength radiation (less energetic) is to the *right*. In astronomy, wavelength is usually measured in Angstroms (A) where 1 A = 10^{-10} m. Visible light ranges from 4,000 to 7,000 A. Note: The SI unit for wavelength is the nanometre where 1 nm = 10 Angstrom or 1 A = 0.1 nm.

visible light. The extreme ultraviolet, abbreviated EUV, is the ultraviolet light closest to x-rays, and is the most energetic of the three types. The far ultraviolet, abbreviated FUV, lies between the near and extreme ultraviolet regions.

X-rays are another form of solar radiation with a very short wavelength. The Sun's x-rays can injure or destroy the tissue of living things. The Earth's atmosphere shields human beings from most of this radiation. Hard x-rays are the highest energy x-rays, while the lower energy x-rays are referred to as soft x-rays. The distinction between hard and soft x-rays is not well defined.

X-rays do not penetrate the Earth's atmosphere. Therefore they must be observed from a platform launched above most of our atmosphere (Fig. 1.14).

The Sun's Magnetic Field

A magnetic field is a region in space in which magnetism exerts a force. The Sun has an overall magnetic field just like a bar magnet – with a north and south pole. Magnetic field lines move outwards from a north pole and inwards at a south pole. The Sun generates its magnetic field internally through the motion of its electrically charged gases (plasma). This field is embedded inside the Sun but it also surrounds the Sun with a 3D shape. At the Sun's poles, the magnetic field is fairly uniform. Nearer the equator the magnetic field gets stretched, amplified and pulled as the Sun rotates because the rotation is not uniform. Zones near the equator rotate with a period of about 26 days, while those near the poles have a period of about 36 days – this effect is called differential rotation.

In 1960 the American astronomer Horace Babcock showed that after several rotations, the magnetic field lines become tangled as they wrap around the Sun. Kinks or loops of magnetic field lines erupt through the surface forming sunspots (the Alpha effect). Babcock's model also help's explain how the polarity of the leading spot in a sunspot group in the northern hemisphere is opposite to those in the southern hemisphere (Hale's law). One problem with Babcock's model is that the expected twisting is far too much and it produces magnetic cycles that are only a couple years in length. More recent dynamo models assume that the twisting is due to the effect of the Sun's rotation on the rising "tubes" of magnetic field from deep within the Sun.

Many scientists now think that the Sun's magnetic field is generated in both the convection zone and in a thin layer called the tachocline, located near the bottom of the convection zone.

Interactions between different magnetic fields on the Sun make the overall magnetic field pattern rather complex. See Fig. 1.15.

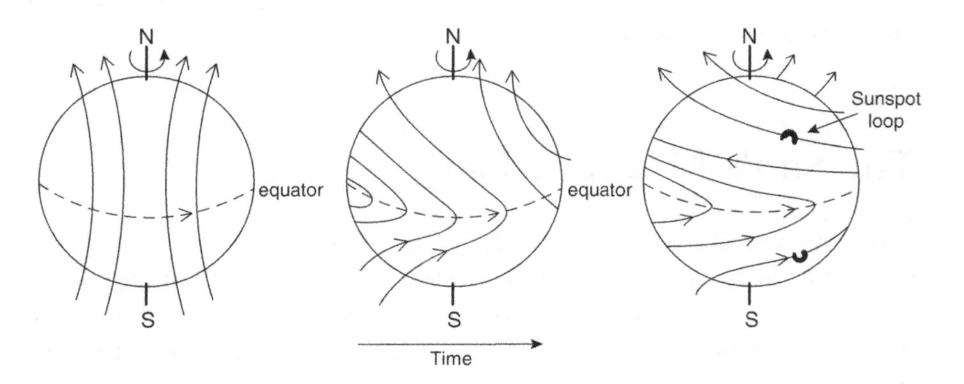

Fig. 1.15 Babcock's Magnetic Dynamo model showing how differential rotation causes the Sun's magnetic field to become twisted as it wraps around the Sun. As a result of this, sunspots form when magnetic loops protrude through the Sun's surface (see third picture).

Did You Know?
American Horace Babcock observed a reversal of the polarity of the Sun's magnetic field during a time of maximum solar activity in 1957–1958. Subsequent investigations showed that the Sun's polarity often undergoes changes. Usually the poles have opposite polarities. It was reported by Pete Riley of the Science Application International Corporation in San Diego, that beginning in March 2000 for nearly a month, the Sun's south magnetic pole faded and the north polarity emerged in its place. Thus during this period, both poles of the Sun had the same polarity. According to Riley, the south pole never really vanished; it simply migrated north and for a while became a band of south magnetic flux, smeared around the Sun's equator. By May 2000 the south magnetic polarity returned to its usual location near the Sun's southern spin axis, but not for long. In 2001 the solar polar magnetic field completely flipped polarity, the south and the north poles then swapped positions. In 2003–2004, the polarity of the north pole was positive; the field lines directed outwards, while in the south polar region, the field lines were directed inwards.

Changes in the Sun's overall magnetic field usually take place in conjunction with the 11-year sunspot cycle. If the polarity of the Sun is N–S in one cycle, then it is S–N in the next cycle. Thus it takes two complete sunspot cycles to return to the original N–S configuration. This means one magnetic cycle has a period of 22 years. At the same time the polarity of the sunspot groups in each hemisphere also change. The preceding polarity spot is usually the dominant "leader" in most groups for the entire 22-year sunspot cycle. The magnetic axis of a sunspot group is usually slightly inclined to the solar east–west line, tilting from 3° near the equator to 11° at latitude 30°N/S, with the preceding polarity spot being slightly closer to the equator (see Fig. 1.16). Understanding the Sun's magnetism is an area of on-going research. Further details in Chap. 3.

The Sun–Earth Relationship

The Sun has a steady output of charged particles and other matter that is collectively known as the solar wind. This wind streams through the Solar System at about 400 km/s. This wind interacts with the atmosphere of Earth and charged particles in particular get trapped in the Earth's magnetic field (the magnetosphere).

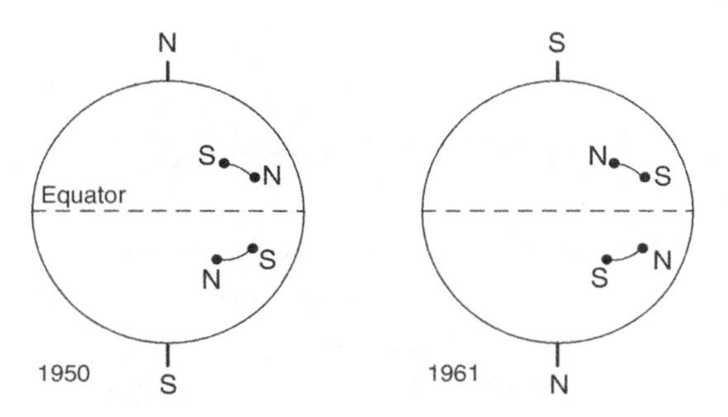

Fig. 1.16 The polarity of the Sun's magnetic field changes with each sunspot cycle (usually every 11 years). The picture shows the polarity in 1950 compared to 1961.

Our magnetic fields and atmosphere in some ways protect us from some solar radiation and cosmic rays from outer space, but they also let in some of the radiation. Outbursts of radiation from the Sun can have dramatic effects on Earth. When a burst of solar radiation strikes the Earth's magnetic field the result can be geomagnetic storms that spark huge electric currents and distort the magnetosphere. This can adversely effect radio communication and navigational systems, and pump extra electricity into power lines (sometimes causing blackouts).

Researchers believe that changes in solar activity may be having an indirect effect on Earth's climate. Satellite measurements, for example, have detected a small change in the Sun's total output during the course of each sunspot cycle. The ebb and flow of solar radiation can heat and cool the atmosphere of Earth enough to change its circulation patterns, which may have significant impacts on regional weather. Researchers have developed powerful computer models to simulate the impact of the Sun on our climate. One such effort, the Whole Atmosphere Community Climate Model (WACCM), helps researchers study interactions among different levels of the atmosphere, ranging from the surface of earth to the upper atmosphere and the edge of space. The modelling work is combined with the analyses of data from observing instruments aboard satellites to track the impacts of solar radiation throughout the atmosphere.

Much of what we have learned has been realized in only the last few decades. Solar space missions such as NASA's TRACE (Transition Region and Explorer Spacecraft) and the SOHO (Solar and Heliospheric Observatory) have provided answers to many questions regarding the effect of the Sun on Earth. But there is a lot of work still to be done and many new questions need answering (see Chap. 9).

Alternative Models of the Sun

The Standard Solar Model treats the Sun as a spherical ball of burning gas in varying states of ionisation with the hydrogen in the deep interior being completely ionised plasma. The Sun is seen to be powered 100% by hydrogen fusion reactions that occur at very high temperatures. This gaseous model is basically the same for all stars. However, some astronomers believe that the standard model fails to explain many of the observed solar phenomena. In one new solar model, it is claimed that the energy to power the Sun came from a supernova core and that the main nuclear reactions involve neutron emission and decay with only 38% hydrogen fusion. The main elements in the Sun are reported to be iron, nickel, oxygen and silicon rather than hydrogen, helium and carbon. Three alternative models that will be briefly discussed below are the Electric Sun model, the Solid Sun model and the Condensed matter model. Each of these alternative models has characteristics in common and it may be that they differ only in name.

The Electric Sun

In the 1970s, Ralph Juergens proposed the Electric Sun model as an alternative to the standard model. In this model, the Sun is surrounded by a plasma cell that stretches far out into space – past Pluto. The Sun is at a more positive potential (voltage) than is the space plasma surrounding it – probably in the order of ten billion volts. Positive ions leave the Sun and electrons enter the Sun. Both of these flows add to form a net positive current leaving the Sun. The model suggests the Sun may be powered not from within itself, but from outside, by the electric currents that flow in our arm of the galaxy. The brightness of the Sun's corona is expected in an electrical model and is due to "normal glow" mode plasma discharge.

The essence of the Electric Sun hypothesis is an analysis of the electrical properties of its photosphere and the chromosphere and the resulting effects on the charged particles that move across them. Electric currents produce magnetic fields and these can be used to explain features like sunspots, coronal holes and flares.

The Solid Surface Model of the Sun

The solid surface model of the Sun proposes that the Sun has a solid, electrically conductive surface composed of iron ferrites beneath the liquid-like plasma layer of the photosphere.

The surface is covered by a series of plasma layers, starting with calcium, silicon, neon, helium and finally a layer of hydrogen that ultimately ignites the corona. The model is based on observations of the Sun by various satellites such as SOHO, TRACE and YOHKOH, and it is able to explain the Sun's behaviour rather well. Supporters of this model claim that images of the Sun obtained from satellites show that the Sun has a uniformly rotating and solid surface below the photosphere. The surface behaves very differently from the photosphere. The surface is being dynamically reshaped and eroded by continual electrical arcing between magnetically polarised points. These arcs emit light consistent with a number of iron ferrites ions, suggesting this surface is composed of ferrite-based materials. The erosion caused by the constant electrical arcing between points along the surface is at least one of the catalysts for the Sun's ever changing sunspots. The electrical discharge process also releases massive amounts of heat into the liquid-like plasma of the photosphere, creating the granular patterns that we see in the photosphere.

The new satellite images of the Sun's surface lend strong support for the electrical model of the Sun.

The Condensed Matter Model of the Sun

Supporters of this model claim that the continuous nature of the Sun's spectrum does not support a gaseous model. This is mainly because gases are known to emit radiation only in discrete bands – they cannot generate thermal spectra in the absence of a rigid body enclosure. Condensed matter can easily generate continuous spectra. It is also claimed that the Sun is relatively incompressible – a characteristic of the liquid state. The Sun has a density of 1.4 g/cm^3 that is closer to that obtained for the condensed state (>1 g/cm^3). The condensed state also better supports seismological findings on the Sun. Doppler images of the solar surface show the presence of transverse waves – something that is unique to the condensed state. In the liquid model, the chromosphere represents that region where matter projected into the corona is in the process of re-condensing in order to enter the liquid state of the photosphere. The presence of superheated liquids within the solar interior can easily explain the production of solar eruptions and the appearance that the surface is boiling. It is also well known that the Sun has strong magnetic fields – on Earth such fields are always produced by condensed matter.

These alternative models have had difficulty gaining momentum and most astronomers are still in favour of the standard gas model. However recent satellite images that allow astronomers to see inside the various layers of the Sun, have provided evidence to renew interest in these models. For further information, do an internet search on "The Electric Sun", "The surface of the Sun" and "The Condensed matter model of the Sun" – also see Web Notes at the end of this chapter.

Web Notes

Further information about the Sun can be found at the following sites:
http://www.solarscience.msfc.nasa.gov
http://www.window2universe.org/sun/atmosphere/corona.html
http://helios.gsfc.nasa.gov/sspot.html
http://www.solarviews.com/eng/sun.htm

For more information about solar neutrinos see Wikipedia "The solar neutrino problem".
For information about alternative models of the Sun see:
http://www.thesurfaceofthesun.com and
http://www.electric-cosmos.org/sun.htm and
http://www.thermalphysics.org/Sun.evidence.1.pdf

2. Probing the Sun

Scientists have gained much of their knowledge about the Sun from observations made on Earth over many years. However, much of our current knowledge has come from space probes that have been sent on missions to observe the Sun. These probes have provided accurate information about the Sun's temperature, atmosphere, composition, magnetic fields, flares, prominences, sunspots and internal dynamics. A knowledge of these probes and the data they collected helps us to better understand the various processes in the Sun and the effect solar radiation has on Earth. This chapter provides information about early solar space probes (pre-1990) as well as the more recent (post-1990) probes and future probes. Satellite and instrumental technology improved greatly during these periods and this is clearly evident when comparing the results obtained from recent probes to the early probes. Table 2.1 lists significant solar space probes.

Early Solar Probes

The USA launched a number of unmanned solar probes between 1959 and 1968 as part of its Pioneer program. Many of these early probes have now completed their missions but some still remain in orbit around the Sun.

Explorer Program

The Explorer program is a USA/NASA program that provided flight opportunities for solar physics and astrophysics investigations from space. The explorer program was the United State's first successful attempt to launch an artificial satellite. The program includes 92 missions since the launch of Explorer 1 in 1958. Besides being the first US satellite, it is known for discovering the Van Allen radiation

J. Wilkinson, *New Eyes on the Sun*, Astronomers' Universe,
DOI 10.1007/978-3-642-22839-1_2,

Table 2.1 Significant solar space probes

Significant solar space probes
Explorer program – USA, Launched many probes since 1958
Pioneer 5 – USA, Launched March, 1960
Pioneer 6 – USA, Launched December 1965 (still transmitting from solar orbit)
Pioneer 7 – USA, Launched August 1966 (recently turned off)
Pioneer 8 – USA, Launched December 1967 (still transmitting from solar orbit)
Pioneer 9 – USA, Launched November 1968 (still in orbit, but died in 1987)
Orbiting Solar Observatory – USA, Launched 1962–1975 (series of nine probes)
Skylab – USA, Launched May 1973 (space station in Earth orbit)
Helios 1 – USA/Germany, Launched November 1974 (came to within 44 million km of the Sun)
Helios 2 – USA/Germany, Launched January 1976 (came to within 43 million km of the Sun)
Solar Maximum Mission – USA, Launched February 1980 (monitored solar flares)
Ulysses – USA/ESA, Launched Oct 1990 (orbited sun's polar regions)
Yohkoh – Japan/USA/UK, launched 31 Aug 1991 (studied high energy radiation from Sun)
SOHO - USA/ESA, launched 2 Dec 1995 (study solar wind, corona, internal structure)
WIND and Polar – launched November 1994 (study solar wind/magnetosphere)
ACE – launched 25 Aug 1997 (study composition of corona/interplanetary space)
TRACE – USA, launched 2 April 1998 (study solar magnetic fields, corona)
Genesis – USA, Launched 8 Aug 2001 (collected solar wind particles and returned them to Earth)
Coronas-F – Russian, Launched July 2001 (monitor flares and solar interior)
RHESSI – USA, Launched February 2002 (x-ray and gamma ray imaging of flares)
Hinode – Japan/USA/UK, Launched 23 September 2006 (explore solar magnetic fields)
Stereo A/B – USA, Two probes launched Oct 2006 (study coronal mass ejections in 3D)
SDO – USA, launched February 2010 (study effects of Sun on Earth)
SOLO – ESA, due for launch 2017 (study Sun from close quarters)
Solar probe plus – USA, due for launch 2018 (study corona and solar wind)

belt around Earth. Explorer satellites have also make important discoveries about the solar wind, solar plasma, solar energetic particles, and atmospheric physics.

Pioneer Probes

Pioneer 5 was launched in March 1960 from Cape Canaveral in the USA. It was a 0.66 m diameter sphere with 1.4 m span across its four solar panels. It was equipped with four scientific instruments: a telescope to detect solar flare particles and observe terrestrial trapped radiation, a magnetometer to measure magnetic field strengths in interplanetary space, a radiation counter to measure cosmic radiation and a micrometeorite detector.

Pioneer 6, 7, 8 and 9 were the first of four identical solar orbiting spacecraft. Pioneer 6, launched in 1965 into solar orbit, is the oldest of NASA's spacecraft believed to be still active. A successful contact was made with Pioneer 6 for about 2 h on December 8, 2000 to commemorate its 35th anniversary. The probe is powered by a 79-W solar panel and consists of an aluminium cylinder 94 cm in diameter and 89 cm long. There were three magnetometer booms, and an antenna mast extending from it.

Pioneers 6–9 demonstrated the practicality of spinning a spacecraft to stabilize it and to simplify control of its orientation. Measurements made by these spacecraft provided much of the early knowledge of the interplanetary environment and the effects of solar activity on Earth. New information was gathered about the solar wind, solar cosmic rays, the structure of the Sun's plasma and magnetic fields, the physics of particles in space, and the nature of storms on the Sun that produce solar flares. Simultaneous measurements by Pioneer 6 and 8 when they were 161 million km apart allowed the most accurate determination of the solar wind density to be made up to that point.

Missions such as Pioneer 10 and 11 showed that gravity assists were possible and that spacecraft could survive high-radiation areas.

Orbiting Solar Observatory

The Orbiting Solar Observatory (OSO) was a series of nine stabilized orbiting platforms developed by the Goddard Space Flight Centre in the USA for observing the Sun and extra solar sources at ultraviolet, x-ray, and gamma-ray wavelengths. NASA launched eight successfully between 1962 (OSO 1) and 1975 (OSO 8) using Delta rockets. Their primary mission was to observe an 11-year sun spot cycle in UV and x-ray spectra. The OSO 9 probe was planned but never launched.

OSO-1 was the first satellite to carry onboard tape recorders for data storage and instruments that could be accurately pointed. Other results of the OSO series included the first full-disc photograph of the solar corona, the first x-ray observations from a spacecraft of a beginning solar flare and of solar streamers and the first observations of the corona in white light and extreme ultraviolet.

Skylab

America's first space station, Skylab, launched in May 1973, was used to study the Sun from Earth orbit. The space station included the Apollo Telescope Mount (ATM), which astronauts used to take more than 150,000 images of the Sun. Solar experiments included photographs of eight solar flares, and produced valuable results that scientists stated would have been impossible to obtain with unmanned spacecraft. The existence of the Sun's coronal holes was confirmed because of these efforts. X-ray photographs taken by Skylab showed that the corona is highly structured, containing coronal loops and holes, and bright x-ray points. However, these images were of poor quality because the telescope had low resolution and low sensitivity (Fig. 2.1).

Skylab was abandoned in February 1974 and re-entered the Earth's atmosphere in 1979. It broke up on re-entry.

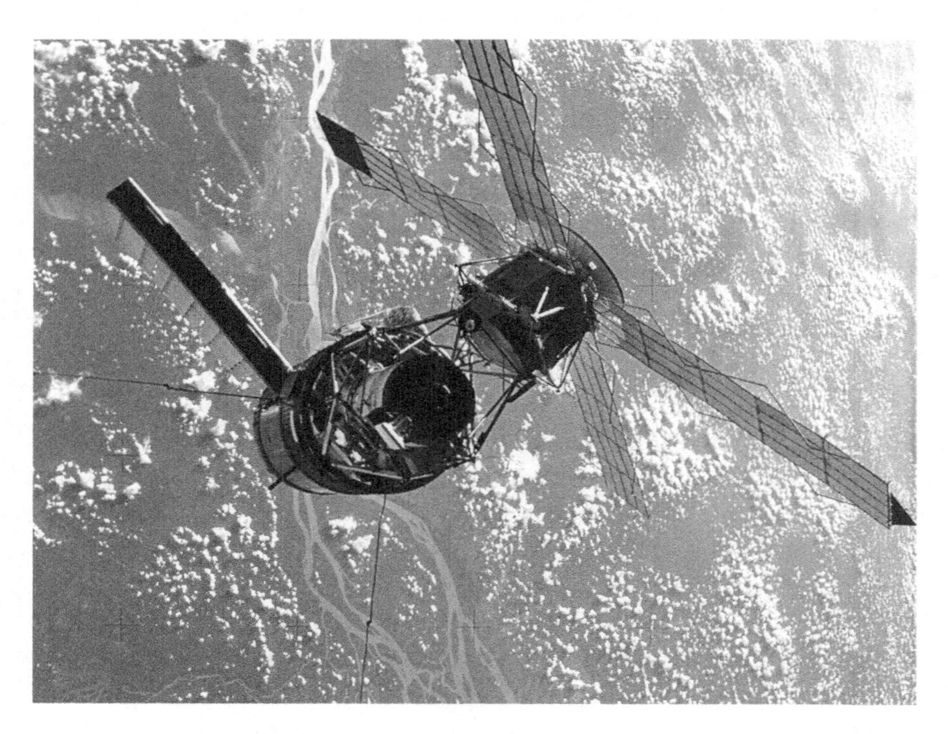

Fig. 2.1 Although it remained in orbit around Earth, the USA's first space station, Skylab, was used to take more than 150,000 images of the Sun. It confirmed the existence of coronal holes on the Sun (Credit: NASA).

Helios 1 and 2

Helios-A and Helios-B (also known as Helios 1 and Helios 2) was a pair of probes launched into orbit around the Sun for the purpose of studying solar processes. A joint venture of the Federal Republic of Germany (West Germany) and NASA, the probes were launched from the John F. Kennedy Space Centre at Cape Canaveral, Florida, on Dec. 10, 1974, and Jan. 15, 1976, respectively.

The probes are notable for having set a maximum speed record among spacecraft at 252,792 km/h. Helios 2 flew to within 44 million km of the Sun (slightly inside the orbit of Mercury). Data was obtained about the velocity and distribution of the solar wind, the intensity of the solar magnetic field and distribution of cosmic rays. Energy transported in the solar wind was found to be carried by protons. Measurements from Helios showed the solar wind has two main velocities. When the wind speed is high, the proton density is relatively low. When the wind speed is low, the proton density is high. In the high-speed wind, heavier particles also have a higher temperature; but it is the other way around in the slow wind, where lighter particles are hotter.

The Helios space probes completed their primary missions by the early 1980s, but they continued to send data up to 1985. The probes are no longer functional but still remain in their elliptical orbit around the Sun. The trajectory of the probes is shown in Fig. 2.2.

Solar Maximum Mission

The Solar Maximum Mission (SMM) was a solar probe launched by the USA on 14th February 1980. The craft was designed to monitor solar flares during a period maximum solar activity. Instruments on the craft were also used to measure solar irradiance. The probe excelled in x-ray and gamma ray spectroscopy of solar flares, as well as making observations of white light emissions from coronal mass ejections. A gamma ray spectrometer was used to detect energetic solar neutrons near the Earth following a solar flare, which occurred on 21 June 1980.

The probe suffered a failure during orbit of the Sun and had to be repaired by space shuttle astronauts in 1984. SMM collected

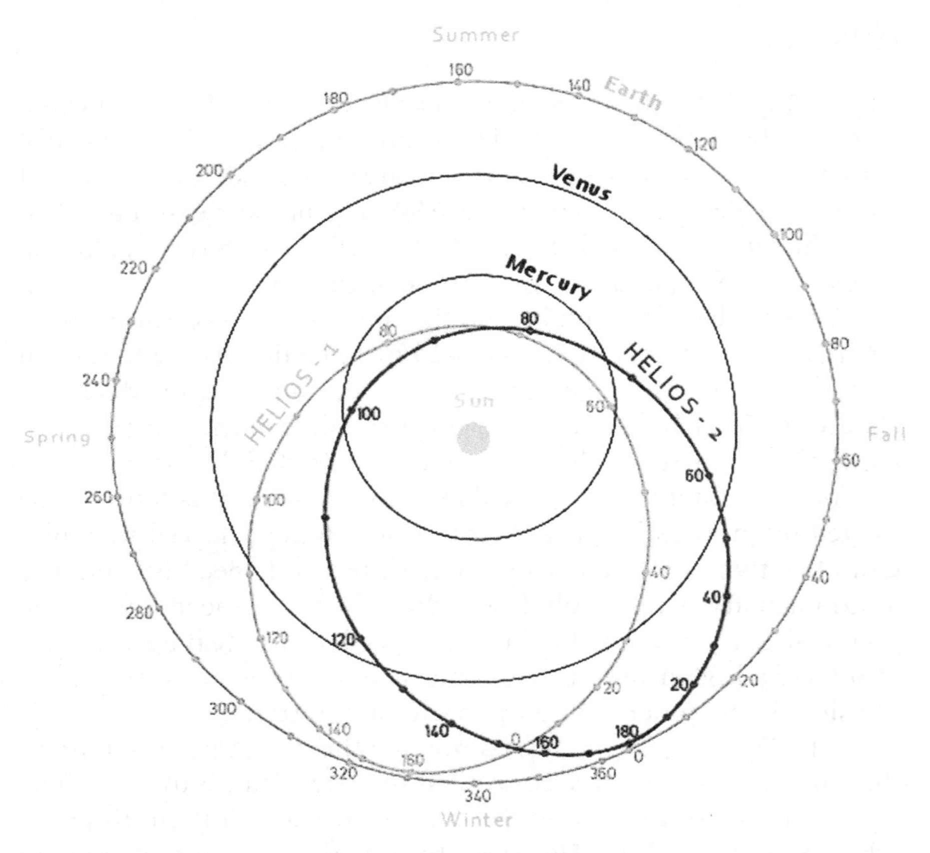

Fig. 2.2 The trajectory of the Helios probes was unusual because they were designed to make measurements in the medium between the inner planets as well around the Sun (Credit: NASA).

data about the Sun until 24th November 1989, and re-entered Earth's atmosphere on 2nd December 1989.

Ulysses

The Ulysses spacecraft, launched on 6th October 1990, was a joint venture between NASA and the European Space Agency (ESA) project designed to study the poles of the Sun and interstellar space around the poles. The first solar polar passage was in June 1994. In February 1995 the craft passed the solar equator in June of that year the craft flew over the Sun's north pole. The spacecraft made a second pass over the Sun's poles between September 2000 and January 2001. At this time, the Sun was close to the peak of its activity cycle. The main objective of the Ulysses mission is to study the properties of the solar wind as a function of latitude, solar magnetic field, solar radio bursts, plasma waves, solar x-rays, solar and galactic cosmic rays. Its instruments found that the solar wind blows faster at the poles than at equatorial regions. Ulysses is now heading back out to the orbit of Jupiter on the long leg of its 6-year circuit around the Sun.

Recent Solar Probes

Yohkoh

Yohkoh means 'sunbeam' in Japanese and is one of the most productive solar space missions conducted by the Institute for Space and Astronautical Sciences in Japan with collaboration with American and British scientists. The spacecraft is in a slightly elliptical low-earth orbit, with an altitude ranging from approximately 570 to 730 km. The orbital period is 90 min.

The scientific objective of Yohkoh is to observe the energetic phenomena taking place on the Sun, specifically solar flares in x-ray and gamma ray emissions.

Yohkoh was launched in August 1991 and contains two spectrometers and two x-ray telescopes (one for hard x-rays and the other for soft x-rays). Each is designed to observe a limited range of wavelengths emitted by the hot plasma produced during solar flares. Observations of spectral lines provided information about the temperature and density of the hot plasma, and about motions of the plasma along the line of sight. Information about temperature and density of the plasma emitting the observed x-rays is obtained by comparing images acquired with different filters. Flare images can be obtained every 2 s. Smaller images with a single filter can be obtained as frequently as once every 0.5 s.

Yohkoh data showed that solar flares and coronal mass ejections can be triggered by magnetic reconnection, where oppositely directed magnetic fields merge together, releasing the necessary energy at the place where they touch. The soft x-ray telescope also showed that the flares or ejections are triggered when the bright x-ray emitting coronal loops become twisted into a helical or S-shape. The probe also found that the corona is ever-changing and has no permanent features.

During each orbit, about five or six times a day, Yohkoh passes over Japan and data is down and up loaded at these times. In addition, Kennedy Space Centre in the USA also receives data from the spacecraft. At other locations in the orbit, the data gets sent to ground stations in the NASA Deep Space network.

Yohkoh collaborated with NASA's High Energy Solar Spectroscopic Imager (HESSI), providing crucial calibration data for its high-resolution hard x-ray images. Solar-B is the Japanese follow-up mission, again with involvement from the US and the UK. It will look at the Sun in soft x-rays, as Yohkoh did, but it will also make very high-resolution images in visible light.

SOHO

The Solar and Heliospheric Observatory (SOHO) another joint NASA/ESA mission launched in December 1995 provided valuable information about the solar atmosphere, solar wind and the Sun's internal structure. SOHO weighs nearly two tonnes and flies in a halo orbit around the Lagrangian point L1. This point is about 1.5 million km from Earth towards the Sun. The L1 Lagrangian point is a location in space where gravitational pull of the Sun and Earth cancel or balance each other. From this vantage point SOHO is able to 'hover' and observe the Sun continuously for 24 h a day, 7 days a week.

In June 1998, ground controllers lost contact with SOHO due to telemetry error. Intense efforts to restore contact paid off 6 weeks later when the spacecraft responded to commands sent from ground stations (Figs. 2.3 and 2.4).

One of SOHO's instruments, called the Large Angle and Spectrometric Coronagraph (LASCO) routinely monitors a huge region of space around the Sun. LASCO is able to take images of the solar corona by blocking the light coming directly from the Sun with an occulter disk, creating an artificial eclipse within the instrument itself. Occasionally, coronal mass ejection can be seen moving away from the Sun. Although not designed for the purpose, the coronagraph instrument on SOHO has detected over

Fig. 2.3 This picture taken prior to launch, shows some of the sophisticated technology on board SOHO (Credit: NASA).

Fig. 2.4 The SOHO space probe contains a total of 12 scientific instruments used to monitor different parts of the Sun (Credit: NASA).

2,000 different comets passing close to the Sun. Of course, it is not SOHO itself that discovers the comets – that is the province of the dozens of amateur astronomer volunteers who daily pore over the fuzzy lights dancing across the pictures produced by the LASCO cameras. Over 70 people representing 18 different countries have helped spot comets over the last 15 years by searching through the publicly available SOHO images online. See Fig. 2.5.

Two other instruments on board the SOHO spacecraft, the Solar Wind Anisotropies (SWAN) and the Michelson Doppler Imager (MDI), allow scientists to 'see' what is happening on the far side of the Sun.

The Extreme-ultraviolet Imaging Telescope (EIT), aboard SOHO takes full-disc images of the Sun's transition region and lower corona at three lines of ionised iron, Fe IX, Fe XII, and Fe XV, and one line of ionised helium, He II. See Table 2.2.

One of SOHO's most important discoveries has been in locating the origin of the solar wind at the corners of honeycomb-shaped magnetic fields near the Sun's poles. Data obtained from SOHO enabled scientists to compare the behaviour of sunspots during low and high activity periods. The space probe has also

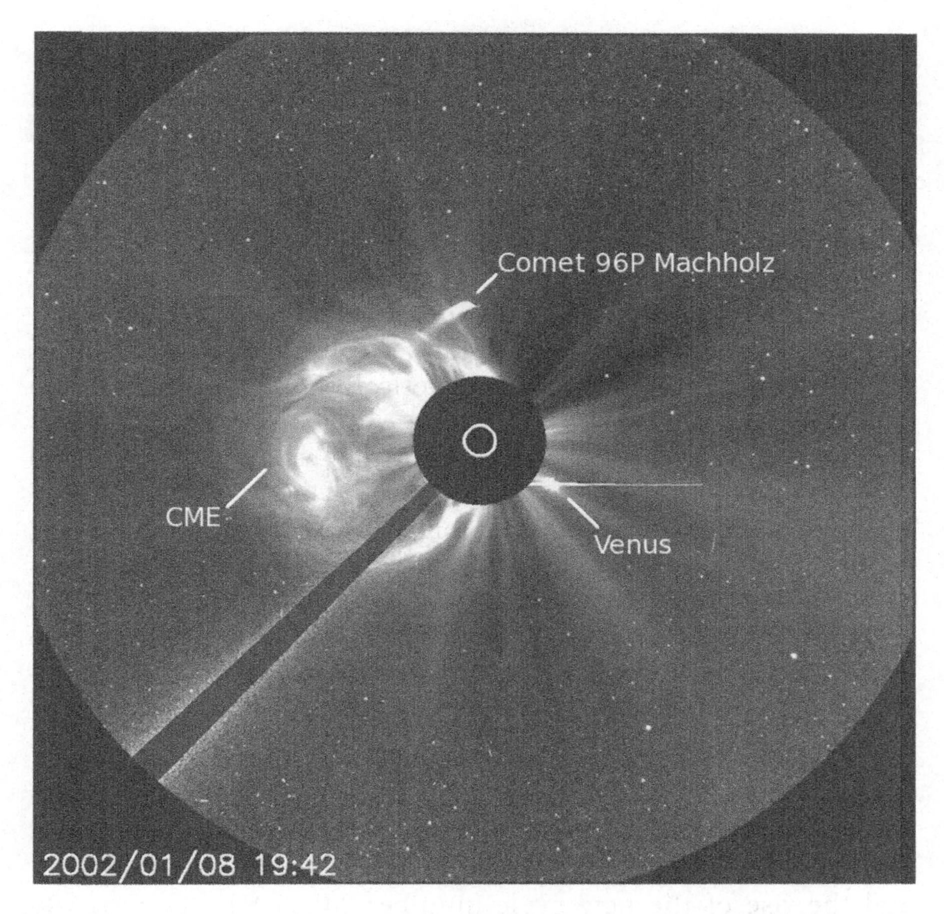

Fig. 2.5 One comet discovered by SOHO is Comet 96P Machholz. The comet orbits the Sun approximately every 6 years and SOHO has seen it three times. Up until January 2011, SOHO had detected over 2,000 comets in orbit around the Sun. A coronal mass ejection (*CME*) and the planet Venus are also shown in this picture (Credit: NASA/ESA/SOHO).

Table 2.2 Wavelengths used by EIT to take images of the Sun's disc

Wavelength (A)	Emitting ion	Formation temperature (°C)
171	Iron, Fe IX and Fe X	1,000,000
195	Iron, Fe XII	1,400,000
284	Iron, Fe XV	2,100,000
304	Helium, He II	60,000

provided an unprecedented breadth and depth of information about the Sun, from its interior, through the hot and dynamic atmosphere, to the solar wind and its interaction with the interstellar medium.

Some of the key findings include:

- Revealing the first images ever of a star's convection zone (its turbulent outer shell) and the structure of sunspots below the surface.
- Providing the most detailed and precise measurements of the temperature structure, interior rotation, and gas flow in the Sun's interior.
- Measuring the acceleration of the slow and fast solar wind.
- Identifying the source regions and acceleration mechanism of the fast solar wind in the magnetically 'open' regions at the Sun's poles.
- Discovering new solar phenomena such as coronal waves and solar tornadoes.
- Revolutionizing our ability to forecast space weather, by giving up to 3 days notice of Earth-directed disturbances, and playing a lead role in the early warning system for space weather.
- Monitoring the total solar irradiance as well as variations in the extreme ultra violet flux, both of which are important to understand the impact of solar variability on Earth's climate.

SOHO was designed for a nominal mission lifetime of 2 years. Because of it's amazing results, the mission has been extended five times (in 1997, 2002, 2006, 2008, and 2010). These extensions allowed SOHO to cover an entire 11-year solar cycle (number 23) and the rise of the new cycle (number 24). SOHO is currently approved through to the end of 2012.

Wind and Polar

Wind and Polar are sister spacecraft used to measure the mass, momentum, energy flows and time variability, throughout the solar wind-magnetosphere-ionosphere system near Earth. Wind was launched in November 1994. It was initially sent into a lunar swing-by orbit in which the Moon's gravity helped propel it through Earth's magnetosphere on its sunward side, at the Lagrange L1 location in space.

Wind observations compliment those of the Polar spacecraft, which looks down at the Earth's magnetic polar regions.

Since launch, Wind has investigated shocks generated by coronal mass ejections, using radio signals to track them from launch in the corona through interplanetary space to the Earth. The most powerful shocks where found to produce a wider wavelength range of radio emission. Radio signals produced by these shocks are triangulated using instruments aboard the Wind, Ulysses, Cassini and twin Stereo spacecraft, permitting a three-dimensional determination of their trajectory.

One of Wind's instruments showed that the slowest solar winds contain the greatest amount of helium, while faster winds have the least amount of helium. Scientists are now using data from Wind to determine where the solar energetic particles originate from, how they are accelerated, and how they escape and propagate from solar flares or coronal mass ejections.

ACE

The Advanced Composition Explorer (ACE) was launched in August 1997 into an orbit around the L1 Langrangian point between the Sun and Earth. The main objective of ACE is to determine the composition of several samples of matter in the solar corona and interplanetary space. It also provides real time solar wind data and monitors solar particles as they bombard Earth. The probe therefore contributes to forecasts of space weather. ACE carries a set of nine instruments to measure the charge state composition of the nuclei from hydrogen to nickel from the solar wind and galactic cosmic rays. One recent discovery of the Wind and ACE spacecraft was of solar wind magnetic fields that merge and join together near the Earth's orbit, in long, steady reconnection layers that stretch out for hundreds of Earth radii. This magnetic reconnection was previously thought to occur only within very small regions in a very short, patchy manner.

TRACE

In April 1998, NASA launched a satellite into sun-synchronous polar orbit around Earth at an altitude of 625 km. The probe called Transition Region And Coronal Explorer (TRACE) is a computerised satellite launched from a Pegasus XL rocket dropped from a jet aeroplane flying high above the Pacific Ocean. The main instrument on TRACE is a Cassegrain telescope of diameter 30 cm and length 160 cm. TRACE is one of several small satellites in NASA's Small Explorer (SMEX) project.

TRACE observes the Sun in extreme-ultraviolet radiation of specific spectral lines sensitive to a wide range of temperatures; but unlike EIT (onboard SOHO), which images the entire solar disc, TRACE observes specific regions on the Sun with higher resolution, detecting fine details that could not be seen from previous spacecraft. The main regions studied by TRACE are the solar photosphere, transition region and corona at temperatures between 4,000 and 5 million °C at wavelengths of 171, 195, 284, 1,216, 1,550, 1,600, 1,700 and 2,500 A.

TRACE is also used to study the connection between the Sun's magnetic field and the heating of its corona. It particular it has provided new insights into the magnetised coronal loops that shape and constrain the hot plasma in solar active regions. The magnetised loops stretch up to 500,000 km from the visible solar disc, spanning up to 40 times the diameter of planet Earth. These magnetic loops also oscillate, or move back and forth, with periods of between 2 and 7 min, appearing when flares excite the oscillations (Fig. 2.6).

TRACE has also provided new information about the quiet corona and transition region, away from solar active regions. Bright, extreme-ultraviolet, flare-like events, known as nanoflares, appear to flash on and off but with insufficient energy to heat the corona. Trace has also observed the jet-like spicules that emerge upwards from the photosphere into the lower corona. More than 10,000 spicules can be seen at any moment on the Sun, rising and falling every 5 min or so, and carrying a mass of 100 times that of the solar wind into the lower corona.

TRACE provided images at five times the magnification of those taken by the Extreme Ultraviolet Imaging Telescope Instrument aboard SOHO. Many details of the fine structure of the corona were observed for the first time. Early in its mission, TRACE discovered the fine-scale magnetic features where enhanced heating occurs at the foot points of coronal loop systems in solar active regions, which later became known as 'coronal moss.'

Fig. 2.6 EUV image taken by TRACE on 21st April 2002, of plasma gas being channelled by magnetic fields into bright, thin loops stretching high into the corona (Credit: NASA/TRACE).

In 2001, TRACE observations of astonishing coronal activity were highlighted in the IMAX movie SolarMax.

Genesis

The Genesis space probe launched by NASA in August 2001 was designed to collect samples of solar wind particles and return them to Earth for analysis. The spacecraft collected solar wind particles from the L1 Langrangian point over about 2.5 years. On its return to Earth in September 2004, a capsule containing the samples was ejected from Genesis but its parachutes did not open and the capsule hit the Utah desert floor at nearly 320 km/h. Several hours after the landing, scientists retrieved the collection canister after the wreckage was made safe. Although the crash left the solar particles open to contamination, scientists were still hopeful of obtaining useful data about the composition of the particles in the solar wind. Scientists wanted a sample from our Sun because a preponderance of evidence suggests that the outer layer of the Sun preserves the composition of the early solar nebula. Knowing the exact elemental and isotopic composition of the outer layer of the Sun is effectively the same as knowing the elemental and isotopic composition of the nebula. The data could help scientists understand how planets and other solar-system objects formed; this would also aid in understanding stellar evolution and the formation of solar systems elsewhere in the universe.

Initial tests showed Genesis collected about 0.4 mg of solar particles, equal only to a few grains of salt. Scientists involved in the research announced on 10th March 2008 that analysis has shown that the Sun has a higher proportion of oxygen-16 than does the Earth. The measurement was made after the upper 20 nm of a collection wafer was removed with a beam of caesium ions. This implies that an unknown process depleted oxygen-16 from the Sun's disk of protoplanetary material prior to the coalescence of dust grains that formed the Earth. See Fig. 2.7.

Fig. 2.7 The Genesis spacecraft was used to collected samples of the solar wind from space. Some samples were collected on its hexagonal wafers and special foils (Credit: NASA/Genesis).

CORONAS-F

The Complex ORbital near earth Observations of the Solar Activity (CORONAS-F) is a Russian space probe launched on 31st July 2001 on a cyclone rocket from Russia's Northern Cosmodrome in Plesetsk. It orbits Earth in a polar orbit at an altitude of 500 km. The main objective is to collect data about solar flares and the solar interior. The craft contains 15 instruments, including ten x-ray spectrometers and imagers, two UV instruments, a radiometer, a coronagraph and several full disc photometers. Being Russian, little information is known about the results from the mission.

RHESSI

The Rematy High Energy Solar Spectroscopic Imager (RHESSI) is a NASA mission aimed at exploring the particle emission and energy release of solar flares. This mission was launched in February 2002. It contains two imaging spectrometers and has the ability to provide high-resolution images of solar flares over a broad spectral range from soft x-rays to gamma rays. It takes full disc solar images. The probe provided scientists with enough data to enable them to calculate far more precisely the exact roundness of the Sun. These measurements indicated that the Sun is not exactly spherical; instead there are small differences between the equatorial and polar radii that result in an oblate shape. The Sun was found to have a thin, rough skin, with bright, magnetic ridges arranged in a network pattern, as on the surface of a cantaloupe.

Hinode (Solar-B)

Hinode (formally Solar-B) is a Japanese Aerospace Exploration Agency solar mission in collaboration with the USA and United Kingdom. It is the follow-up mission to the Yohkoh and it was launched from Japan on 23rd September 2006 and placed in a sun-synchronous orbit.

Hinode was planned as a 3-year mission to investigate the solar magnetic fields and their role in heating the chromosphere and corona. It consists of a 50 cm solar optical telescope and spectrophotometer, an extreme ultraviolet (EUV) spectrometer, and an x-ray telescope.

Hinode's x-ray telescope has provided new information about the energy source of the Sun's corona. It discovered twisted and tangled magnetic fields that are able to store huge amounts of energy. When the complicated magnetic structures relax to simpler configurations, a huge amount of energy is released. This energy heats the corona and powers solar eruptions like flares and coronal mass ejections. The x-ray telescope also discovered gigantic arcing magnetic structures surrounding the active regions of sunspots.

How the solar wind is formed and powered has been the subject of debate for decades. Data from Hinode also showed that powerful magnetic 'alfven' waves play a critical role in driving the solar wind into space. In the past, alfven waves have not been able to be seen because of limited resolution in available instruments. With the help of Hinode, scientists have been able to see direct evidence of alfven waves, which will help them unravel the mystery of how the solar wind is powered.

Unlike instruments on TRACE and other dedicated solar observatories, the x-ray telescope on Hinode, is a 'grazing incidence' telescope capable of studying so-called 'soft' x-rays. Most solar telescopes observe lower-energy radiation. The x-ray telescope also photographs the Sun faster and with better resolution and greater sensitivity than previous grazing-incidence x-ray instruments.

The Hinode solar optical telescope was the first to be able to measure small changes in the Sun's magnetic field. The data collected will be used to study how these changes evolve and coincide with dynamic events seen in the corona.

The EUV imaging spectrometer has been designed to measure the flow velocity or speed of solar particles, and diagnose the temperature and density of solar plasma. The EUV imager provides a crucial link between the other two instruments because it can measure the layers that separate the photosphere from the corona.

STEREO A/B

The Solar TErrestrial RElations Observatory (STEREO) is a solar mission launched by NASA on 26th October 2006. It consists of two nearly identical spacecraft, one orbiting ahead of Earth (A) and the other behind Earth (B). Observations are made simultaneously of the Sun and then combined to provide a 3D stereo image of the Sun. Spacecraft A takes 347 days to orbit the Sun while spacecraft B takes 387 days. Because the A spacecraft is moving faster than B, they are separating from each other and A is orbiting closer to the Sun than B. The images are adjusted to account for this difference.

Each of the spacecraft carries cameras (a EUV imager and two coronagraphs), particle experiments and radio detectors in four instrument packages. STEREO is used to image the inner and outer corona and the space between Sun and Earth, detect electrons and other energetic particles in the solar wind, study the plasma characteristics of protons, alpha particles and heavy ions, and monitor radio wave disturbances between the Sun and Earth (Fig. 2.8).

From February 2011, the two Stereo spacecraft will be 180° apart from each other, allowing the entire Sun to be seen for the first time. Such observations will continue for several years. By combining images from the STEREO A and B spacecraft, with images from NASA's Solar Dynamic Observatory (SDO) satellite, a complete map of the Sun can be formed. Previous to the STEREO mission, astronomers could only see the side of the Sun facing Earth, and had little knowledge of what happened to solar features after they rotated out of view. In 2015 contact with the two spacecraft will be temporarily lost for a few months as they both pass behind the Sun. After this, they will continue to operate again and approach Earth.

SDO

The Solar Dynamics Observatory (SDO) is the most advanced spacecraft ever designed to study the Sun and its dynamic behavior. SDO is providing better quality, more comprehensive science data faster than any NASA spacecraft currently studying the Sun. The probe is aimed at providing data on the processes inside the

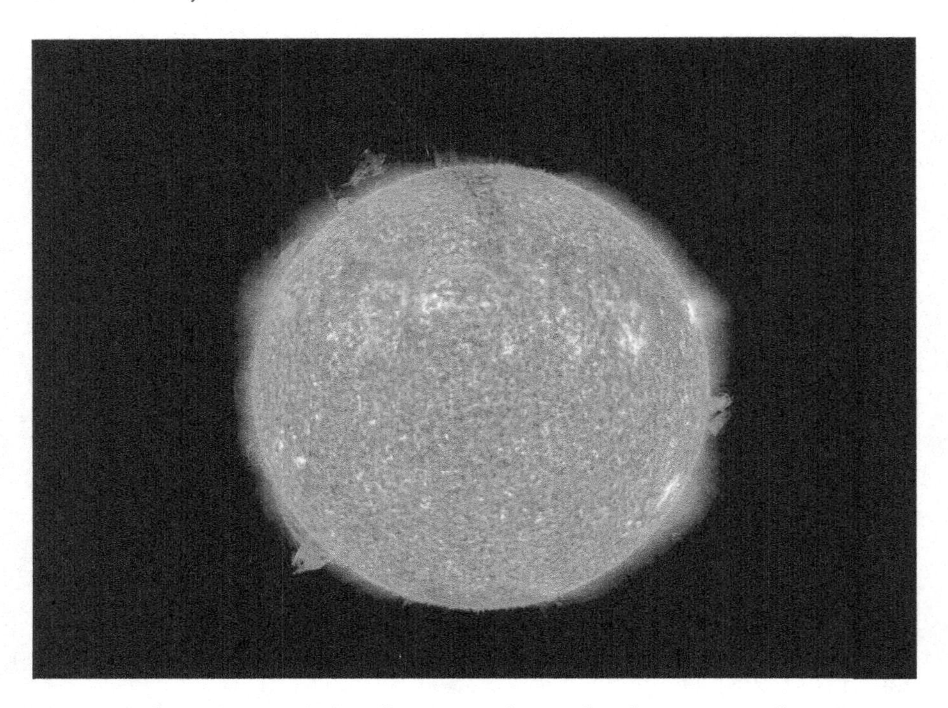

Fig. 2.8 The STEREO (Ahead) spacecraft caught this spectacular eruptive prominence in extreme UV light as it blasted away from the Sun (12–13th April 2010). This was certainly among the largest prominence eruptions seen by either the STEREO or SOHO missions. The length of the prominence appears to stretch almost halfway across the Sun, about 800,000 km. Prominences are cooler clouds of plasma that hover above the Sun's surface, tethered by magnetic forces. They are notoriously unstable and commonly erupt as this one did in a dramatic fashion (Credit: NASA/STEREO).

sun, the sun's surface, and its corona that result in solar variability. SDO will help scientists to better understand the Sun's influence on Earth and near-Earth space through the use of many wavelengths simultaneously. SDO will also investigate how the Sun's magnetic field is generated and structured.

SDO was launched from Cape Canaveral Air Force Station in the USA on 11th February 2010. After launch, SDO was placed into an orbit around Earth at about 2,500 km. It then underwent a series of orbit-raising maneuvers that placed it in a circular, geosynchronous orbit at altitude 36,000 km. It has a 5-year science mission and carries enough fuel to operate for an additional 5 years. At launch its mass was 3,100 kg with a payload of 290 and 1,450 kg of fuel. The solar panels cover an area of 6.6 m^2 producing

Fig. 2.9 The Solar Dynamics Observatory (*SDO*) is the most advanced spacecraft ever designed to study the Sun (Credit: NASA/SDO).

1,450 W of power. The overall length of the spacecraft along the Sun-pointing axis is 4.5 m, and each side is 2.22 m. See Fig. 2.9.

The SDO has three main instrument packages:

- The Atmospheric Imaging Assembly (AIA) is an array of four telescopes that observes the surface and atmosphere of the sun. The AIA filters cover ten different wavelength bands that are selected to reveal key aspects of solar activity. For example, wavelengths of 1,600 and 1,700 A are used to monitor the photosphere and transition regions; 304 A is used to monitor the chromosphere, while 171 and 193 A are used to monitor the corona.
- The Extreme Ultraviolet Variability Experiment (EVE) measures fluctuations in the Sun's ultraviolet output. Extreme ultraviolet (EUV) radiation from the Sun has a direct and powerful effect on Earth's upper atmosphere; it provides heat and inflation, and inserts enough energy to break apart atoms and molecules.
- The Helioseismic and Magnetic Imager (HMI) maps solar magnetic fields and peers beneath the Sun's opaque surface using a technique called helioseismology. A key goal of this experiment is to decipher the physics of the Sun's magnetic dynamo.

Did You Know?

The rapid cadence and continuous coverage required for Solar Dynamics Observatory (SDO) observations led to placing it into an inclined geosynchronous orbit. This allows for a nearly continuous, high data-rate contact with a single, dedicated ground station. Nearly continuous observations of the Sun can be obtained from other orbits, such as low Earth orbit. If SDO were placed into a lower orbit, it would be necessary to store large volumes of scientific data onboard until a downlink opportunity was available, and multiple sites around the globe would be needed to downlink the data. However, no space-qualified data recorder with the capability to handle this large data volume exists. This lack of a data recorder, the large data rate of SDO, and the ability to continuously stream data from the spacecraft if a geosynchronous orbit was selected, led to the selection of the inclined geosynchronous orbit. The disadvantage of this inclined geosynchronous orbit includes higher launch and orbit acquisition costs and eclipse (Earth shadow) seasons twice annually. During these 2–3 week eclipse periods, SDO experiences a daily interruption of solar observations, and these interruptions have been included in SDO's data capture budget. There will also be three lunar shadow events each year from this orbit.

The inclined geosynchronous orbit is located on the outer edges of Earth's radiation belt, where the radiation dose can be quite high. Additional shielding was added to reduce the effects of exposure to this ionizing radiation. Because the potential for damage due to space radiation effects is a 'space weather' effect, SDO is affected by the very processes it is designed to study (Fig. 2.10).

Fig. 2.10 On 1st August 2010, almost the entire Earth-facing side of the Sun erupted. There was a huge solar flare with multiple filaments of magnetism lifting off the Sun's surface, and large-scale shaking of the solar corona, radio bursts, a coronal mass ejection and more. This extreme ultraviolet snapshot from the Solar Dynamics Observatory (*SDO*) shows the Sun's northern hemisphere in mid-eruption. Different colours in the image represent different gas temperatures ranging from 1 to 2 million °C (Credit: NASA/SDO).

In May 2010, the AIA instrument on the SDO observed a number of very small flares that generate magnetic instabilities and waves over a large fraction of the Sun's surface. The instrument is capturing full disc images in eight different temperature bands that span 10,000 to 36 million °C. This allows scientists to observe entire events that are very difficult to discern by looking in a single temperature band, at a slower rate, or over a more limited field of view.

The data from SDO is providing a lot of new information and spectacular images of the Sun. Scientists are gaining a better understanding of how even small events on the Sun can significantly effect the operation of technological infrastructure on Earth (such as GPS systems, cable TV, radio and satellite communications).

Future Solar Probes

SOLO

The SOLar Orbiter (SOLO) is a solar mission proposed by the European Space Agency (ESA). The mission is planned for launch in January 2017. SOLO will perform detailed measurements of the inner heliosphere and solar wind, and perform close observations of the polar regions of the Sun. At its closest point, the spacecraft will be closer to the Sun than any previous spacecraft (one fifth the distance between Earth and the Sun or within 60 solar radii). It will be able to almost match the Sun's rotation around its axis for several days, and so will be able to see solar storms building up over an extended period from the same viewpoint. It will also deliver data of the side of the Sun not visible from Earth.

SOLO is specifically designed to always point to the Sun, and so its Sun-facing side is protected by a sunshield. The spacecraft will also be kept cool by special radiators, which will dissipate excess heat into space.

SOLO will carry a number of highly sophisticated instruments:

- Solar wind analyser – to measure solar wind properties and composition.

- Energetic particle detector – to measure charged and energetic particles.
- Magnetometer – to measure magnetic fields using high resolution.
- Radio and plasma wave analyser – to measure magnetic and electric fields.
- Polarimetric and Helioseismic imager – to provide high-resolution images of the photospheric magnetic field.
- EUV imager – to image various layers of solar atmosphere.
- EUV spectral imager – to examine surface and corona.
- X-ray spectrometer – to image Sun using x-rays.
- Coronagraph – to image the corona.
- Heliospheric imager – to image flow of solar wind.

SOLO will take about 3 years to reach the Sun using gravity assists from Venus and Earth. These swing-bys will put the spacecraft into a 168 daylong orbit around the Sun from which it will begin its scientific mission. During the course of the mission, additional Venus gravity assist manoeuvres will be used to increase the inclination of SOLO's orbit, helping the instruments to see the polar regions of the Sun clearly for the first time. Solar orbiter will eventually see the poles from an angle higher than 30°, compared to 7° at best from Earth.

Solar Probe Plus

Solar probe plus is planned for launch by NASA by July 2018. On its 6-year mission, the probe will perform 24 close-pass manoeuvres, using several 'slingshots' around Venus to get progressively closer to the Sun. At its closest approach, the probe will be within six million km of the Sun. To protect it from the Sun's heat, the probe will be tucked away behind a 2.7 m diameter, 15 cm thick shield made from a carbon foam composite; that will withstand over 1,400°C and intense radiation. The probes closest approach will be around December 2024.

The aims of the mission are to:

- Determine the structure and dynamics of the magnetic fields at the sources of solar wind.

- Trace the flow of energy that heats the corona and accelerates the solar wind.
- Determine what mechanisms accelerate and transport energetic particles.
- Explore dusty plasma near the sun and its influence on solar wind and energetic particle formation.

Did You Know?

A number of spacecraft used to monitor the Sun make use of a spectroscope or spectrograph.

Spectroscopes traditionally consist of a prism and several lenses that magnify the spectrum so that it can be examined. After photography was invented, scientists preferred to produce a permanent photographic record of spectra. A similar device for photographing a spectrum is called a spectrograph.

In its basic form a spectrograph consists of a slit, two lenses, and a prism arranged to focus the spectrum of an astronomical object, such as the Sun or a star, onto a photographic plate. See Fig. 2.11.

The spectrograph mounts at the focal point of a telescope and the image of the object being examined is focused on the slit. After the spectrum has been photographed, the spectrum is compared to the spectrum of known elements. Each element produces its own unique set of spectral lines. This method allows scientists to determine what elements are present on the object being examined. Scientists can also determine the velocities of objects from any shift in spectral line wavelengths (the Doppler shift).

A better device for breaking light into a spectrum is the diffraction grating – this device replaces a prism (see Fig. 2.12). In recent years charge coupled devices (CCDs) connected to a computer have replaced photographic plates to record spectra. CCDs produce a spectral graph that plots light intensity against wavelength. Dark spectral lines appear as dips in the graph, while bright lines appear as peaks. CCDs can be used for both visible and UV light. The exact choice of detector depends on the wavelengths of light to be recorded.

The forthcoming James Webb Space Telescope will contain both a near-infrared spectrograph (NIRSpec) and a mid-infrared spectrometer (MIRI).

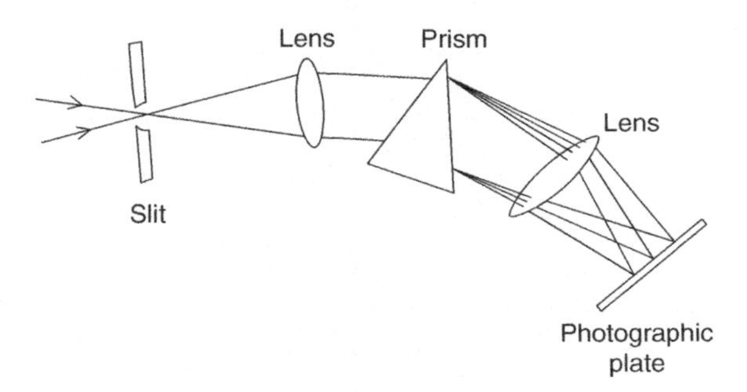

Fig. 2.11 A prism spectrograph.

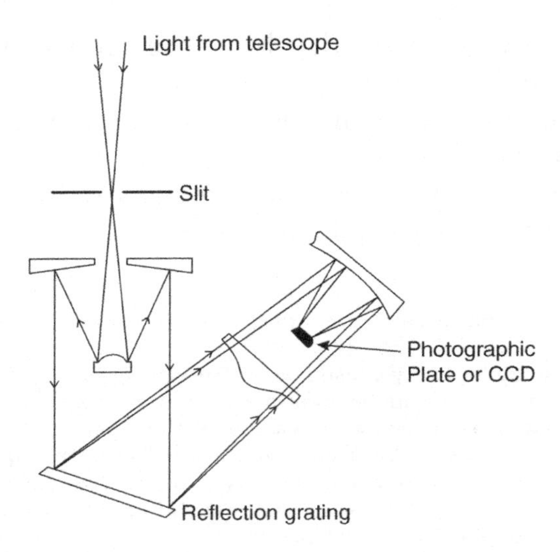

Fig. 2.12 A grating spectrograph.

Web Notes

For information on the Helios space probes see: http://www.honeysucklecreek.net/dss44/helios.html
For information on the Ulysses space probe see: http://ulysses.jpl.nasa.gov/
For information on SOHO see: http://sohowww.nascom.nasa.gov
For information about Hinode see: http://www.isas.jaxa.jp/e/enterp/missions/hinode/
For information on the SDO see: http://sdo.gasfc.nasa.gov
For information on SOLO see: http://sci.esa.int/solarorbiter
For information on Solar probe plus see: http://solarprobe.gsfc.nasa.gov/

3. Activity in the Photosphere

The visible surface of the Sun is a layer called the **photosphere**. This layer is a thin shell of hot, ionised gases or plasma about 400 km thick. Most of the energy radiated by the Sun passes through this layer. From Earth, the surface looks smooth, but it is actually turbulent and granular because of convection currents.

A number of features can be observed in the photosphere with special telescopes. These features include the granules, dark sunspots, and the bright faculae. Scientists can also measure the flow of material in the photosphere using the Doppler effect that involves monitoring changes to the wavelength of lines in the Sun's spectrum. These measurements reveal additional features such as supergranules as well as large-scale flows and a pattern of waves and oscillations.

This chapter will focus on activity in the photosphere of the Sun.

Granulation

Pictures of the Sun's surface reveal a mottled appearance called **granulation**. It is a bit like looking at the surface of an orange. One of the first persons to record such an appearance of the Sun's surface was Sir William Herschel (1738–1822) in 1801. Father Secchi (1818–1878) also described the Sun's surface as being covered by bright grain-like features (granules), separated by dark lanes. The French astronomer Pierre Jules Jannssen (1824–1907) took the first photographs of the granules using a 13.5 cm aperture telescope in 1894. Interest in granulation was revived in 1950 when Richardson and Schwarzchild of Mount Wilson Observatory in the USA made a spectral study of granulation. Their results showed a weak correlation between brightness and velocity fluctuations in the granules. In the 1930s German astronomers

J. Wilkinson, *New Eyes on the Sun*, Astronomers' Universe,
DOI 10.1007/978-3-642-22839-1_3, © Springer-Verlag Berlin Heidelberg 2012

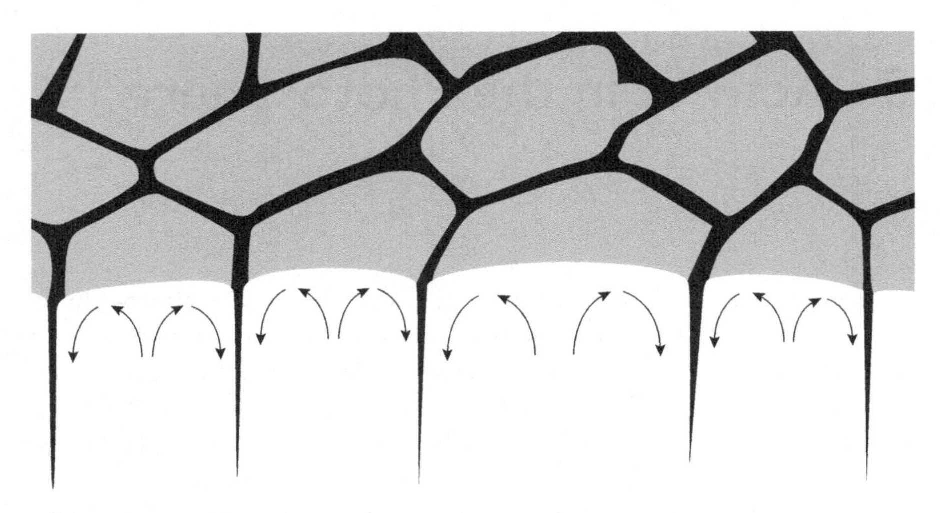

Fig. 3.1 Granules on the Sun's surface are around 1,000 km in diameter. There are at least a million granules visible on the Sun's surface at any moment. They are constantly evolving and changing, producing a honeycomb pattern of rising and falling gas that is in constant turmoil.

Unsold and Siedentopf suggested the granules were due to convection currents in the layers of the Sun below the surface. See Fig. 3.1 and the photo in Fig. 3.2.

It is now well established that granules are convection cells in the Sun's photosphere. By measuring changes in the wavelengths of spectral lines in various parts of the granules, astronomers have found that hot gas rises upward (at about 1 km/s) in the centre of a granule. As it cools, the gas radiates its energy, in the form of visible and electromagnetic radiation out into space. The cooled gas then spills over the edges of the granules and plunges back into the Sun along the boundaries or lanes between granules. A granule is brighter at its centre than at its edges because the centre is at a higher temperature.

Photographs of granules show them to be bright with an irregular, cellular shape. The dark lanes between them are generally of uniform width. The latest high-resolution photographs show individual granules have considerable diversity in their brightness. Bright points appear side by side in dark lanes between granules. These bright patches are believed to be associated with magnetic field concentrations on the Sun and are about 80 km in diameter.

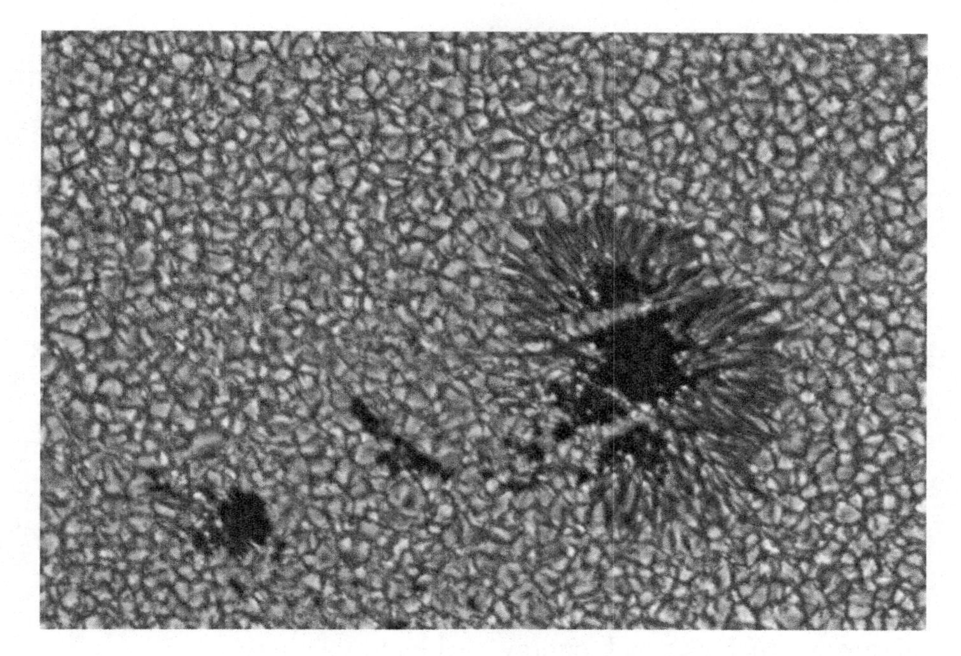

Fig. 3.2 Granules surface of the Sun and a sunspot group (Credit: NASA/GSFC).

Granules seem to increase in size once they form, until they reach a certain size, then they break up into smaller granules, eventually fade and vanish. Most researchers seem to agree that all granules develop from previous granular fragments. The mean lifetime of granules ranges from 6 to 16 min, depending on the size of the granules.

Granules are best seen in solar telescopes with apertures over 80 mm (3 in.).

Supergranulation

The Michelson Doppler Imager (MDI) aboard the SOHO spacecraft has been used to observe supergranulation on the Sun's surface. Such massive cells are thought to be caused by large-scale convection currents below the photosphere. About 3,000 supergranules are seen on the surface of the Sun at any moment. The supercells are between 12,000 and 20,000 km in diameter and last between 16 and 20 h. A typical supercell contains about 900 granules. Material in a supergranule flows out and sideways from its centre, with a typical velocity of 0.4 km/s. After this horizontal motion, the material eventually sinks down again at the cell boundary. The supergranular flow carries the magnetic fields with it.

Supergranulation cannot be seen on white light photographs of the Sun, except near the limb where it is seen as faculae – it can however be seen on images from space probes.

Sunspots

The surface of the Sun contains dark areas called **sunspots**. These spots appear dark because they are cooler than the surrounding photosphere – about 3,500°C compared to 5,500°C of the photosphere. If a sunspot were isolated from the photosphere it would be brighter than an electric arc – but because its temperature is lower than the surroundings it appear dark. Sunspots also radiate only about one fifth as much energy as the photosphere.

Sunspots vary in size from 1,000 km to over 40,000 km and they change shape as they move slowly across the surface of the rotating Sun from east to west. Their lifetime seems to depend on their size, with small spots lasting only several hours, and larger spots or groups persisting for weeks or months.

In 1610, English astronomer, Thomas Harriot was one of the first to observe sunspots through a telescope. Frisian astronomers Johannes and David Fabricius, published a description of sunspots in June 1611. Galileo also observed sunspots around this time (Figs. 3.3 and 3.4).

Observations of the spots as they moved slowly across the face of the Sun can be used to estimate the rotational period of the Sun. Galileo for example, reported that the Sun rotates once in about 4 weeks. Since a typical sunspot group lasts for no longer than 2 months, it can only be followed for up to two rotations of the Sun. The equatorial regions of the Sun actually rotate faster than the polar regions. A sunspot near the equator takes 25 days to go once around the Sun, but a sunspot nearer the poles takes up to 35 days.

In 1700 Edward Maunder noticed that the Sun had changed from a period of high sunspot activity to one of low activity.

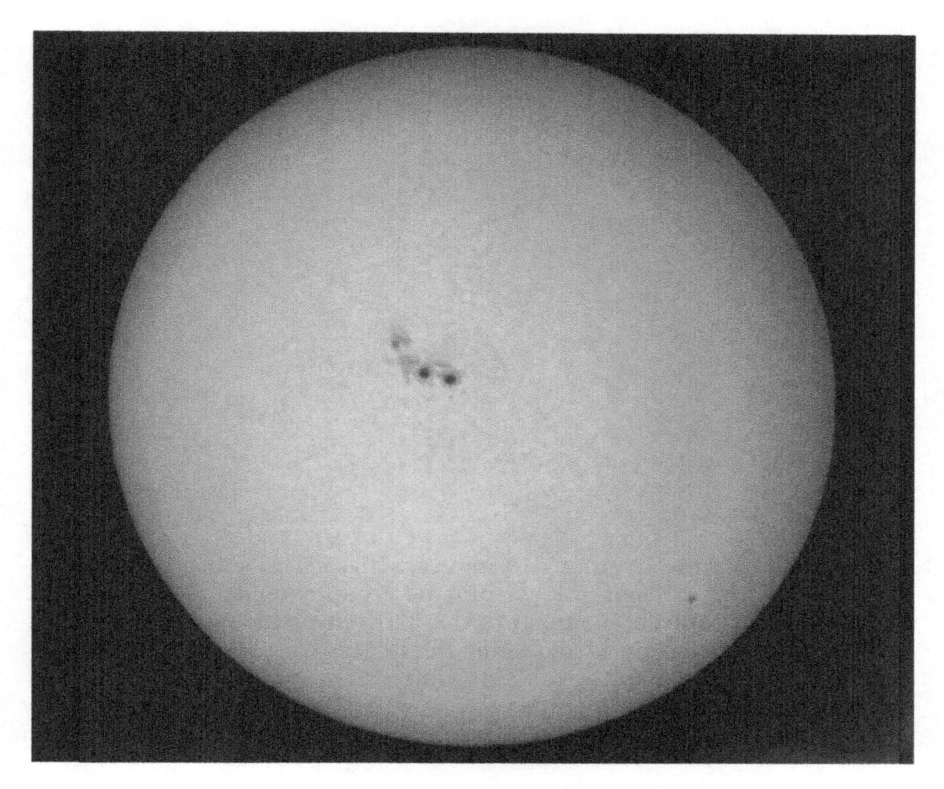

Fig. 3.3 Sunspots on the Sun's surface (Credit: NASA/MSFC).

Sunspots were rarely seen during the second part of the seventeenth century in a time called the Maunder Minimum (1645–1717). This period of solar inactivity also corresponds to a climatic period of cooling on Earth called the "Little Ice Age". A cyclic variation of the number of sunspots was first observed by Heinrich Schwabe between 1826 and 1843 and led Rudolf Wolf to make systematic observations starting in 1848. The Wolf number is a measure of individual spots and spot groupings. In 1848, Joseph Henry reported that sunspots appeared dark because they were cooler than the surrounding surface.

We now know that the number of sunspots on the Sun varies during a cycle lasting about 11 years. A time of many sunspots is called a sunspot maximum. Sunspot maxima occurred in 1968, 1979 and 1990. During a sunspot minimum, the Sun is almost devoid of sunspots, as it was in 1965, 1976 and 1986. This regular variation is known as the 11-year **sunspot cycle**. See Fig. 3.5.

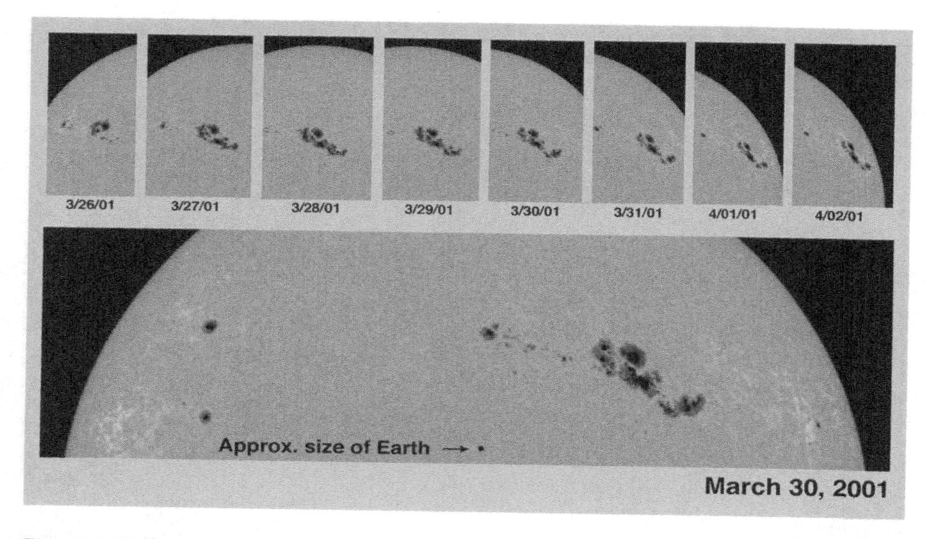

Fig. 3.4 A large sunspot group as it moved with the Sun's rotation. On 30th March 2001, the sunspot area within the group extended across an area more than 13 times the diameter of the Earth. It yielded numerous flares and coronal mass ejections, including the largest x-ray flare recorded in 25 years on 2nd April 2001 (the *top right image* in the series) (Credit: NASA/SOHO).

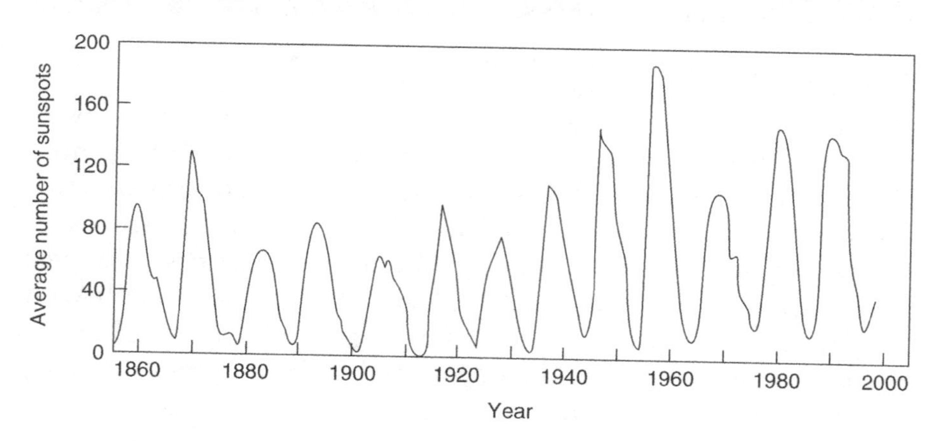

Fig. 3.5 Graph showing how the average number of sunspots for each year has varied since 1850. The cycle of maximum and minimums has an 11-year period.

At the start of each new sunspot cycle, the Sun's overall magnetic field reverses (north becomes south and south becomes north). As a result, the polarity in sunspot groups change in each

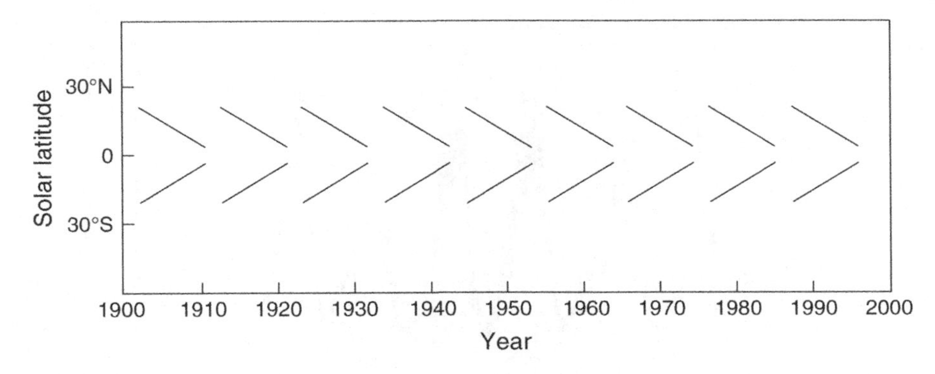

Fig. 3.6 The average latitude of sunspots varies throughout the 11 year Sunspot cycle. This plot of the changing positions of sunspots resembles the wings of a butterfly, and has been called the Butterfly diagram. The vast majority of sunspots occur between the equator and latitude zone 35° north and 35° south.

hemisphere. In order for the magnetic polarities to return to their original state, two solar cycles will have to occur. This reversal and return of the Sun's magnetic field to its original state is called the **magnetic cycle** and it averages 22 years.

The average latitude of new sunspots also changes throughout the sunspot cycle. At the start of each cycle, the sunspots are mostly at about 25° north and south of the equator. Ones that form later in the cycle occur closer to the equator. Figure 3.6 shows how the average latitude at which sunspots are located varies at the same 11-year rate, as does the number of sunspots.

Did You Know?

In May 2009 a team of international scientists sponsored by NASA released a new prediction for the next solar cycle. Solar cycle 23 ended in 2009, and cycle 24 begun. Cycle 24 is expected to peak in May 2013 with a below average number of sunspots. Only about 90 spots are expected during the peak of cycle 24, the lowest of any cycle since 1928 when solar cycle 16 peaked at 78.

Although the peak will be low, there is still the possibility of severe space weather. The great geomagnetic storm of 1859, for instance, occurred during a solar cycle of about the same size as expected for 2013. See Fig. 3.7.

Predicting sunspot numbers in the past has not been easy. At first glance it looks like a regular pattern, but predicting the peaks and valleys has proved troublesome. Cycles vary in length from about 9–14 years. Some peaks are high, others low. The valleys are usually brief, lasting only a couple of years, but sometimes they stretch out much longer. In the seventeenth century the Maunder Minimum lasted for about 70 years.

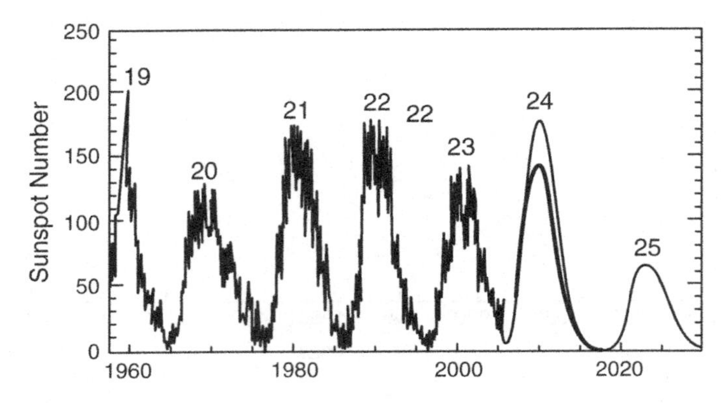

Fig. 3.7 Sunspot numbers since 1960 with predicted numbers for solar cycle 24 and 25.

During 2010, solar cycle 24 was rising, with small spots popping up with increasing frequency. Enormous currents of plasma on the Sun's surface were gaining strength and slowly drifting towards the Sun's equator. Radio astronomers detected a small but significant increase in solar radio emissions. All these things were precursors of solar cycle 24 and formed the basis of a new forecast.

Low solar activity has a profound effect on Earth's atmosphere, allowing it to cool and contract. The becalmed solar wind whips up fewer magnetic storms around Earth's poles.

The Sunspot Number

Scientists track solar cycles by counting the number of sunspots at a particular time. Counting sunspots is not as straightforward as it sounds. If you looked at the Sun through a small, safely filtered telescope, you might be able to see two or three large spots. An observer using a high-powered telescope might see 10 or 20 spots. A powerful space-based observatory could see 50–100. Which is the correct sunspot number?

There are two official sunspot numbers in common use. The first, the daily "Boulder Sunspot Number," is computed by the NOAA Space Environment Centre using a formula devised by Rudolph Wolf in 1848. This formula is given as:

$$R = k\,(10g + s),$$

where R is the sunspot number; g is the number of sunspot groups on the solar disk; s is the total number of individual spots in all the groups; and k is a variable scaling factor (usually <1) that accounts for observing conditions and the type of telescope. Scientists combine data from lots of observatories, each with its own k factor, to arrive at a daily value.

The Boulder number is usually about 25% higher than the second official index, the "International Sunspot Number," published daily by the Solar Influences Data Centre in Belgium. Both the Boulder and the International numbers are calculated from the same basic formula, but they incorporate data from different observatories.

As a simple rule, if you divide either of the official sunspot numbers by 15, you'll get the approximate number of individual sunspots visible on the solar disc as seen by projecting its image on a paper screen with a small telescope.

Sunspots and Magnetic Fields

In 1908 the American astronomer George Hale discovered that sunspots are directly linked to intense magnetic fields on the Sun. Hale noticed that the spectral lines of sunspot light were split into two or more lines – this effect called the **Zeeman effect**, is caused by magnetic fields. The darker the sunspot the more intense the magnetic field and the more the spectral lines are separated. Magnetic fields on the Sun are about 10,000 times stronger than those on Earth's surface.

It is now known that sunspots are areas where a bundle of concentrated magnetic fields project through the hot gases or plasma of the photosphere. The protruding magnetic fields prevent hot gases inside the Sun from rising to the surface as they normally do. Such regions of the photosphere are left relatively devoid of gas and are cooler and darker than the surrounding solar surface.

Hale also found that sunspots usually exist in pairs or bipolar groups, with each member of a pair having opposite magnetic polarity (+ and −). From cycle to cycle, the polarities change from +/− to −/+ and back again. Furthermore, the order of polarity in sunspot pairs in the northern hemisphere is always opposite to those of sunspot pairs in the southern hemisphere. According to the Hale-Nicholson Rules, the preceding polarity spot is usually the dominant "leader" in most groups for the entire 22-year

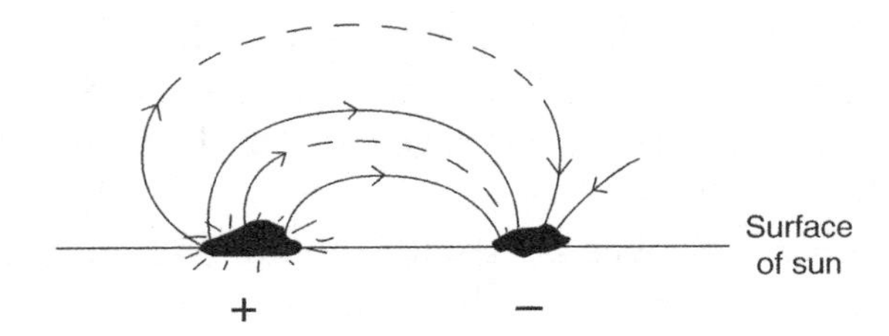

Fig. 3.8 In a sunspot pair, magnetic field lines move up and out of the surface (+ polarity) from one member and down and into the surface (– polarity) at the other member.

sunspot cycle. The magnetic axis of a sunspot group is usually slightly inclined to the solar east–west line (Joy's Law), tilting from 3° near the equator to 11° at latitude 30°N/S, with the preceding polarity spot being slightly closer to the equator. See Fig. 1.16.

Recent observations from the SOHO spacecraft using sound waves travelling below the Sun's photosphere have been used to develop a 3D image of the internal structure below sunspots. These observations show that there is a powerful downdraft underneath each sunspot, forming a rotating vortex that concentrates the magnetic field so as to prevent the up-flow of energy from the hot interior. Sunspots can thus be thought of as self-perpetuating storms, analogous in some ways to hurricanes on Earth. These spots reach down to an average depth of 5,000 km (Fig. 3.8).

The Structure of Sunspots

Each sunspot consists of two parts:

- The central umbra, which is the darkest part, where the magnetic field rises vertically out of the Sun's surface, and
- The surrounding penumbra, which is lighter, where the magnetic field is more inclined to the surface.

Sunspots begin life as a small pore that usually develops an umbra and then a penumbra. Umbrae appear black or a deep reddish-brown colour against the bright background of the Sun's surface. Penumbras are grey in colour and contain bright filaments that radiate about the umbra like fine threads. These filaments contain magnetic fields and are associated with flowing gas. The darker the area in a sunspot, the more intense the magnetic field.

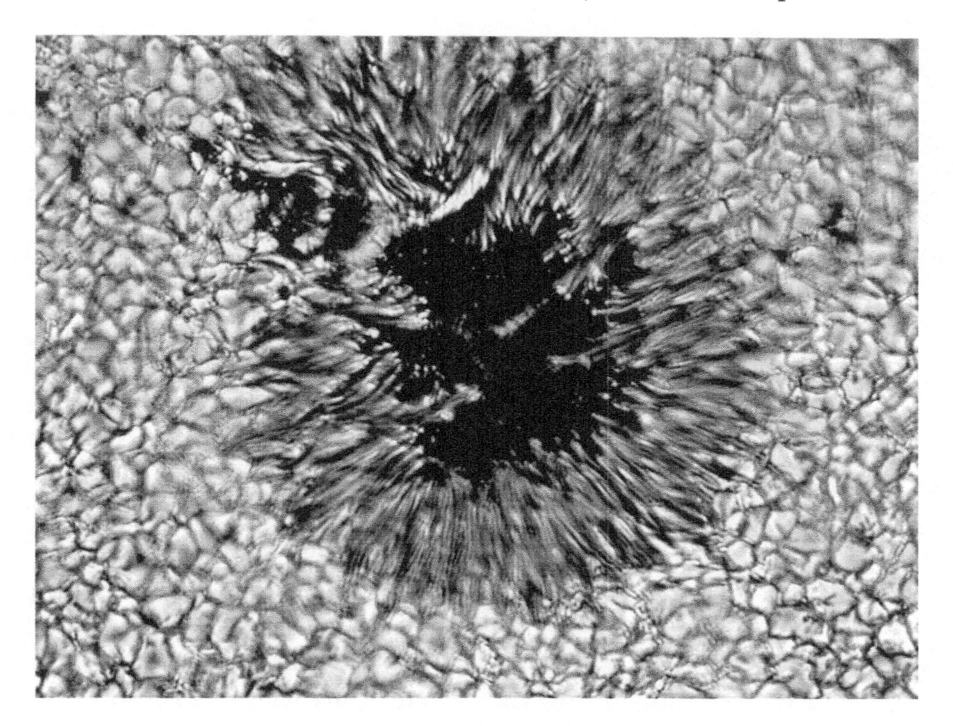

Fig. 3.9 A mature sunspot. The darker central region is the umbra (strong magnetic field). The lighter out region is the penumbra with its filamentary structure (weaker magnetic field). Surrounding the penumbra is the granular photosphere (Credit: NASA, NSO, NOAO).

Once a sunspot forms it usually grows in size over several days as it moves across the Sun's surface. Larger spots are darker and cooler than smaller ones (Figs. 3.9 and 3.10).

New high-resolution pictures from solar space probes show that the filaments in sunspots have dark cores within them. The filaments observed are typically 150–180 km wide, and their dark cores are less than 90 km wide. The discovery of the filament cores will be a valuable aid in studying penumbra, which are difficult to understand, being made up of complex magnetic-field geometry interacting with gas flows and oscillations.

Pictures of the chromosphere in H-alpha light show small, elongated dark line structures called **fibrils**. These filament-like structures tend to run along magnetic field lines. Often, they are connected to or part of the structure of larger filaments, curving into or running along the filament's main axis.

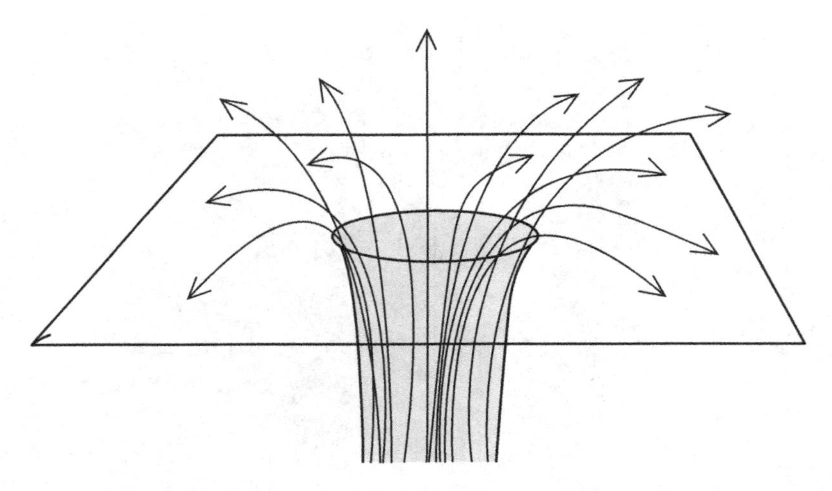

Fig. 3.10 Shape of magnetic field lines (*grey*) in and around a sunspot. The central core contains a dense bundle of magnetic field lines that rise near vertically out of the surface. The penumbra contains less dense magnetic bundles that are more inclined to the surface. Some bundles are pulled down below the surface and some run along the surface.

On white light photographs of sunspots, umbrae often show complex structures called **light bridges**. Scientists have measured the heights of such structures to be 200–450 km above the umbra. Light bridges often show a sunspot is in the final stages of its life. Sometimes an irregular sunspot is divided into two or more smaller spots as a result of a light bridge. Two light bridges can be seen in the umbra of the photo in Fig. 3.2.

Scientists also believe that magnetic structures like sunspots hold the key to space weather. Such weather, originating in the Sun, can affect Earth's climate and environment. A bad storm can disrupt power grids and communication, destroy satellites and even expose airline pilots, crew and passengers to radiation.

Observing Sunspots

Sunspots can be observed with land-based and Earth-orbiting solar telescopes. These telescopes use special filters and projection techniques for direct observation. Specialized tools such as spectroscopes and spectrohelioscopes are also used to examine sunspots and sunspot areas.

It is dangerous to look directly at the Sun with the naked eye as the bright light and heat permanently damages vision. Amateur observation of sunspots is generally conducted indirectly using projected images, or directly through protective filters. A telescope eyepiece can project the image, without filtration, onto a white screen where it can be viewed indirectly, and even traced, to follow sunspot evolution. Special purpose hydrogen-alpha narrow bandpass filters as well as aluminium coated glass attenuation filters (which have the appearance of mirrors due to their extremely high optical density) on the front of a telescope provide safe observation through the eyepiece. Amateur observers need to be extremely careful when using filters to view the Sun. Under no circumstances should you use "dark glass" eyepiece filters labelled as "Sun Filters" – they will crack under the Sun's intense light and heat. See Chap. 6 for further information.

Solar Rotation Effects

As the Sun rotates on its axis, sunspots appear to move across the solar disc from east to west. During this motion the appearance of sunspots changes from an oblong – elongated shape near the east limb, to a more rounded shape near the centre of the disc, to oblong again near the western limb. These changes are due to foreshortening effects as the Sun is a spherical body and the spots are embedded on its surface. See Figs. 3.4 and 3.11.

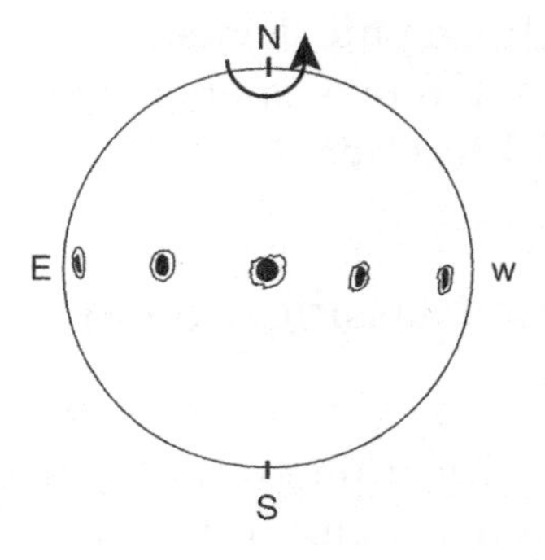

Fig. 3.11 Variation in the appearance of a sunspot as it traverses the solar disc.

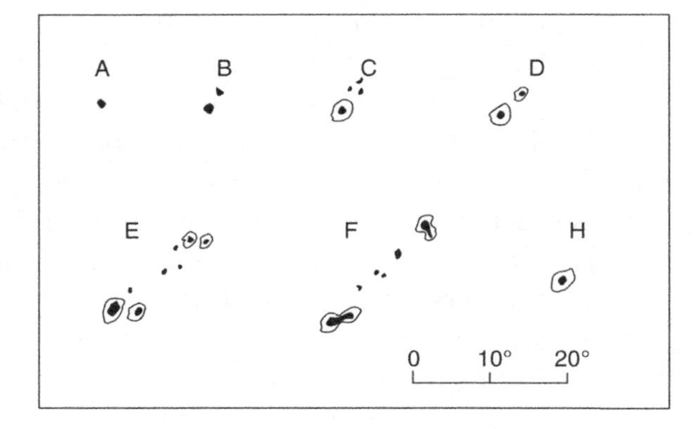

Fig. 3.12 Zurich modified classes for sunspots. The scale is in degrees (remember there are 90° from the centre of the Sun to the limb in each N, S, E, W direction).

McIntosh Sunspot Classification System

Scientists have been classifying sunspots for many years and in many ways. One method of classification, the Zurich Method of Sunspot Classification, was devised in 1938 by M. Waldmeier (Fig. 3.12). An extended version of this scheme was developed by Patrick McIntosh in the 1960s and 1970s. The McIntosh classification system is presented in this book as the most useful for amateur observers (see Table 3.1). Spots are classed according to a three-letter system, for example, Eso, Dai, or Hax.

To understand the system you need to know that:

- A unipolar group is a single spot or a single compact cluster of spots with the greatest distance between two spots of the cluster not exceeding three heliographic degrees.
- A bipolar group contains two spots or a cluster of many spots extending roughly east–west with the major axis exceeding a length of three heliographic degrees.
- A particular spot or group of spots may go through a number of categories in their lifetime.

Mt Wilson Magnetic Classification System of Sunspot Groups

Another sunspot classification system is the Mt Wilson (or Hale) magnetic classification system. In this system, suffixes p or f are used when the preceding or following polarity spot, respectively, is dominant.

Table 3.1 McIntosh sunspot classification system

Modified Zurich class (first letter)
A – A single spot or group of spots that are unipolar and have no penumbra
B – A group of spots that are bipolar and have no penumbra
C – A group of spots that are bipolar and have spots with penumbrae – usually on only one side of an elongated group
D – A group of spots that are bipolar that have spots with penumbrae on both sides of an elongated group. The group length is less than 10° of heliographic longitude
E – A group of spots that are bipolar that have spots with penumbrae on both sides of an elongated group. The group length is between 10° and 15° of heliographic longitude
F – A group of spots that are bipolar that have spots with penumbrae on both sides of an elongated group. The group length of greater than 15° heliographic longitude
H – A single spot or group of spots that is unipolar with a penumbrae
Largest spot in group (second letter, small case, indicates appearance of the penumbra)
x – No penumbrae
r – Incomplete penumbra surrounds only part of spot
s – Symmetric penumbra, <2.5° in N–S diameter
a – Asymmetric penumbra, <2.5° in N–S diameter
h – Symmetric penumbra, >2.5° in N–S diameter
k – Asymmetric penumbra, >2.5° in N–S diameter
Distribution of spots within the group (third letter, small case)
x – Unipolar group, class A or H
o – Open distribution. Very few or none, tiny spots between leader and follower
i – Intermediate. Many spots between leader and follower, none with mature penumbra
c – Compact distribution. Many spots between leader and follower, at least one with mature penumbra

Alpha: A single dominant spot, often linked with a plage of opposite magnetic polarity.
Beta: A pair of dominant spots of opposite polarity (Bipolar, i.e. a leader and a follower).
Gamma: Complex groups with irregular distribution of polarities.
Beta–Gamma: Bipolar groups which have more than one clear north–south polarity inversion line.
Delta: Umbrae of opposite polarity together in a single penumbra.

Just over half of the observed groups are Beta-p or Alpha-p, with the larger groups most often being Beta-p, Beta-Gamma, or Delta.

Delta groups are generally very active and often are the site of major solar flares. They have a complex, irregular, or "broken" umbral look and rarely last for more than one solar rotation.

For amateurs, this system is not attractive as measurements of the magnetic field cannot be made directly.

Magnetograms

The Michelson Doppler Imager on board the SOHO space probe measures the velocity of oscillations produced by sounds trapped inside the Sun, and obtains high-resolution magnetograms of the Sun's surface. The magnetograms indicate that there is a lot of magnetism in the photosphere outside sunspots. The majority of these magnetic fields are concentrated into intense magnetic flux tubes that appear, disappear, and are renewed in just 40 h. The individual flux tubes are a few hundred kilometres across and have magnetic field strengths comparable to those of the much larger sunspots.

The magnetograms also indicate that at sunspot minimum there are still many small areas of magnetism scattered across the surface, even though there are few sunspots.

Magnetograms are available in the image archives of the SOHO website (see Fig. 3.13), but as from the start of 2011 have been replaced by the magnetograms from SDO. See Chap. 7 for further information.

Faculae, Flocculi and Plage

Faculae are patches of great brightness, usually greater than that of the photosphere on the Sun's surface. They are often seen close to the solar limb, where they stand out in marked contrast to the reduced brightness of the photosphere. There shape sometimes changes in the course of a few hours although they may remain for weeks in the same region of the Sun. Faculae have significantly longer lifetimes than sunspots and cover a larger fraction of the solar disc.

Faculae are visible only in white light. They are usually precursor to sunspot groups and are found in the same zones as sunspots. There are no sunspots without faculae, but there are often extensive areas of faculae without spots. Faculae frequently

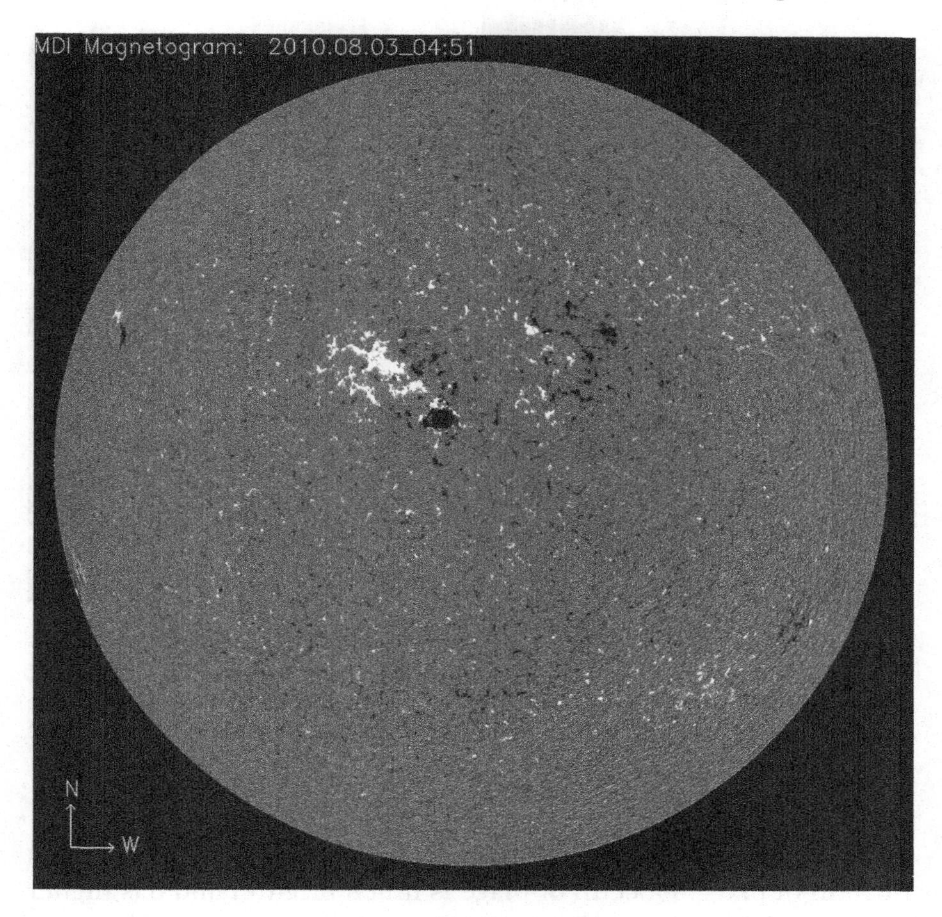

Fig. 3.13 A solar magnetogram taken by SOHO on 3rd August 2010. It shows the solar disc as *grey*. The MDI (Michelson Doppler Imager) picture was taken in the continuum near the Ni I 6768 A line. The most prominent feature is the sunspot seen as the *large black spot in the centre* of the image. The magnetogram image shows the magnetic field in the solar photosphere, with *black* and *white* indicating opposite polarities. Notice that the image appears *flat* (Credit: SOHO/NASA).

act as a connecting link between successive sunspot groups and they tend to cluster around the preceding and the following member of a bipolar sunspot group.

Polar faculae are sometimes seen during a solar cycle minimum but they differ from normal faculae in that they appear as small, short lived, speaks. See Fig. 3.14.

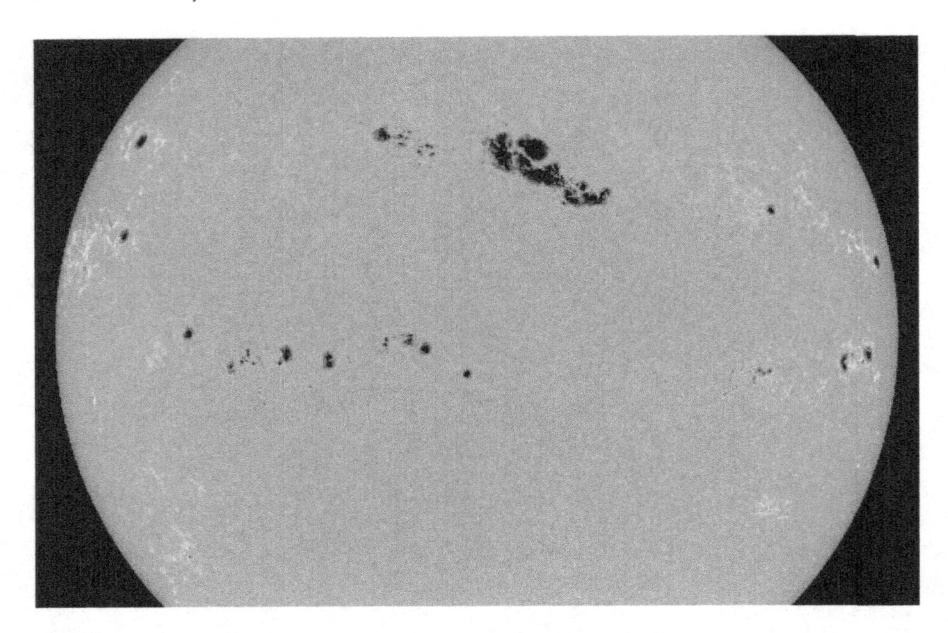

Fig. 3.14 Although sunspots cause a decrease in solar irradiance they are accompanied by bright white blotches called faculae that cause an overall increase in irradiance. On 30th March 2001, the sunspot area within the large group spanned an area more than 13 times the entire surface of the Earth (Credit: NASA/SOHO).

Faculae that exists in the chromosphere is better known as flocculi or plage. Flocculi or plage is much brighter and has higher contrast than photospheric faculae. Sometimes small round flocculi may appear where there was no sign of previous activity. Once flocculi form, the general tendency is for it to stretch and expand with time, towards the equator in both hemispheres. After reaching a maximum area, it tends to break apart into irregular shapes. Sunspots usually always have flocculi but sometimes flocculi exist without sunspots. Flocculi can surround a sunspot as a cloud-like form with no particular consistency in brightness or shape. It is important not to confuse flocculi or plage with a solar flare – a flare is more intense and does not last as long.

Faculae are best observed using white light and direct observation through a telescope with special solar filters rather than via eyepiece projection onto a white screen. Flocculi or plage glows in hydrogen alpha light so a H-alpha solar telescope shows them well.

Did You Know?
For more than two centuries, scientists have wondered how much heat and light the Sun releases, and whether this energy varies enough to change Earth's climate. Around 1976 Jack Eddy, a solar astronomer from the USA, reported that he had found a connection between century-long solar activity and major climatic shifts. In 1978, NASA launched a series of satellites with instruments called radiometers, which measure the amount of sunlight striking the top of the Earth's atmosphere (called total irradiance). Radiometers enabled scientists to determine how much energy the Sun emits and whether it varies with the sunspot cycle. Furthermore they were able to study the competition between dark sunspots and the bright faculae that drive solar irradiance.

Up until the 1970s, scientists assumed that the Sun's irradiance was unchanging and the amount of energy it expelled was even called the "solar constant". At the end of the 1970s scientists realised that the Sun's irradiance fluctuates constantly in conjunction with the sunspot cycle. But the matter is not that simple.

Radiometer measurements collected from space probes during the 1980s and 1990s showed scientists that irradiance is actually a balance between darkening from sunspots and brightening from accompanying hot regions called faculae. When solar activity increases, as it does every 11 years or so, both sunspot and faculae become more numerous. But during the peak of a cycle, the faculae brighten more than sunspots dim it. Overall, radiometers show that the Sun's irradiance changes by about 0.1% as the number of sunspots varies from about 20 spots or less per year during a solar minimum to between 100 and 150 during periods of a solar maximum.

Some scientists believe this 0.1% variation is too low to explain all of the recent global warming on Earth.

Combining results from a number of radiometers into one data stream for long-term study has proven complicated because many radiometers record slightly different absolute measurements and actually degrade while in orbit. Complicating the issue further, an instrument aboard NASA's SOHO probe measured irradiance levels during a solar minimum in 2008 that were actually lower than the previous solar minimum.

A recent NASA probe called Glory contained a radiometer that was more technologically advanced and stable than older models. Glory was due be launched from the USA on 23rd February 2011 in a polar orbit around Earth. It was hoped that results from Glory would provide answers to the solar irradiance problem. However, the rocket carrying Glory crashed shortly after take-off into the Pacific Ocean on March 4, 2011. Glory was a remote-sensing Earth-orbiting observatory designed to achieve two primary mission objectives. One was to collect data on the physical and chemical properties as well as the spatial and temporal distributions of aerosols. The other was to continue collection of total solar irradiance data for the long-term climate record.

It is hoped that a replacement for Glory might be launched in the future.

Web Notes

For information on photospheric features try:
http://solarscience.msfc.nasa.gov/feature1.shtml
For information on sunspots try:

http://sohowww.nascom.nasa.gov/sunspots/ or http://www.windows2universe.org/sun/
or http://solarscience.msfc.nasa.gov/SunspotCycle
For further information on sunspot classification, including sample photos with classes, see: http://sidc.oma.be/educational/classification.php. Click on the highlighted examples to see the pictures.
For information about solar cycle 24 predictions:
http://solarscience.msfc.nasa.gov/predict.shtml

4. Activity in the Chromosphere and Corona

Extending out from the photosphere is the Sun's atmosphere. It can be divided into two main regions – the chromosphere and the corona. The **chromosphere** is a layer of tenuous gas with a density much less than that of the photosphere. It is about 2,500 km thick with a temperature that varies from 6,000°C just above the photosphere to about 20,000–30,000°C at its top. The chromosphere is characterised by spikes of gas called **spicules**, giant flames called **prominences** and **solar flares**. The chromosphere is important to scientists because it is largely responsible for the deep ultraviolet radiation that bathes the Earth, producing our atmosphere's ozone layer, and it has the strongest solar connection to climate variability.

Surrounding the chromosphere is the outer region of the Sun's atmosphere, the **corona**. It extends several million kilometres from the top of the chromosphere into the space around the Sun. There is no well-defined upper boundary to the corona. The corona is the home of the solar wind, coronal holes and coronal mass ejections. Temperatures in the corona reach as high as one million degrees Celsius because of interactions between hot ionised gases and the photosphere's strong magnetic fields.

This chapter will focus on solar activity in the chromosphere and corona.

Spicules

Spicules are one of the most common features within the chromosphere. They are long thin jets of luminous gas, projecting upwards from the photosphere. Spicules rise to the top of the chromosphere

J. Wilkinson, *New Eyes on the Sun*, Astronomers' Universe,
DOI 10.1007/978-3-642-22839-1_4, © Springer-Verlag Berlin Heidelberg 2012

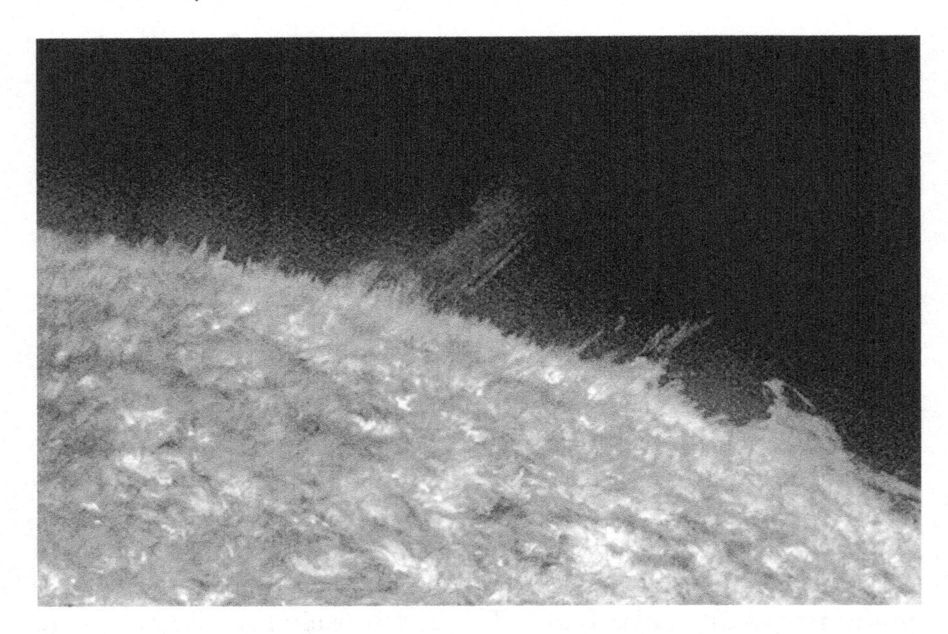

Fig. 4.1 High-resolution images of the chromosphere reveal numerous spikes, which are jets of gas called spicules. A typical spicule rises for 5 min at a supersonic rate of 72,000 km/h to a height of nearly 10,000 km (Credit: NASA/SDO).

at about 72,000 km/h and then sink back down again over the course of about 10 min. They consist largely of ionised gas that follows magnetic field lines. See Fig. 4.1.

Spicules seen along the edge of the Sun are fuzzy and somewhat hairy in appearance. They rise to a height of nearly 10,000 km and have a diameter around 800 km. Each spicule lasts for about 5 min, before falling back into the Sun or disappearing from view. New spicules continually form as old ones fade away. They are generally located on the boundaries of enormous regions of rising and falling gas called **supergranules**.

Groups of spicules take on different shapes and patterns – clusters of spicules are sometimes called 'bushes' while columns or rows are called a 'chains' (Fig. 4.2).

Scientists using images from the SOHO, TRACE and Hinode satellites together with the Swedish Solar Telescope have been able to produce a more complete picture of the gas inside these giant fountains. Tracking the movement and temperature of spicules relies on successfully identifying the same phenomenon

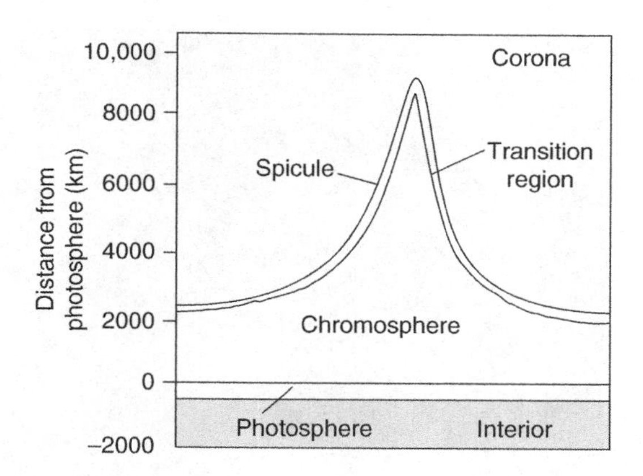

Fig. 4.2 Spicules are jets of hot plasma gas that rise upwards into the chromosphere. The jets last for only a few minutes before falling back into the Sun or disappearing.

in the images from each source. One complication comes from the fact that different instruments 'see' gas at different temperatures and in different wavelengths. Pictures from Hinode in the visible light range, for example, show only cool gas, while pictures that record UV light show gas that is up to several million degrees.

Scientists have recently found that spicules are driven by sound waves generated within the Sun. These sound waves have the same 5 min oscillation periods as the spicules. Although the sound waves usually are damped, under certain conditions, they burst into the chromosphere generating shock waves. These waves propel the plasma upwards as tall tubes of gas, producing more than 100,000 spicules at any given time on the Sun's surface (Fig. 4.3).

In 2012 NASA is scheduled to launch the Interface Region Imaging Spectrograph (IRIS), which will focus on the density, temperature and magnetic field between the surface of the sun and the corona. Researchers are hoping that data from this mission will lead to a better understanding of spicules and coronal heating.

Did You Know?

One of the most puzzling mysteries in solar physics is why the Sun's outer atmosphere, the corona, is millions of degrees hotter than its surface. In early 2011, scientists from Lockhead Martin's Solar and Astrophysics Laboratory, the National Centre for Atmospheric Research, and the University of Oslo, announced they had found that jets of hot plasma (spicules) shooting up from the Sun's surface might be responsible for heating the corona.

For decades scientists believed spicules could send heat into the corona. However, following observational research in the 1980s, it was found that spicule plasma did not reach coronal temperatures, and so the theory was discarded.

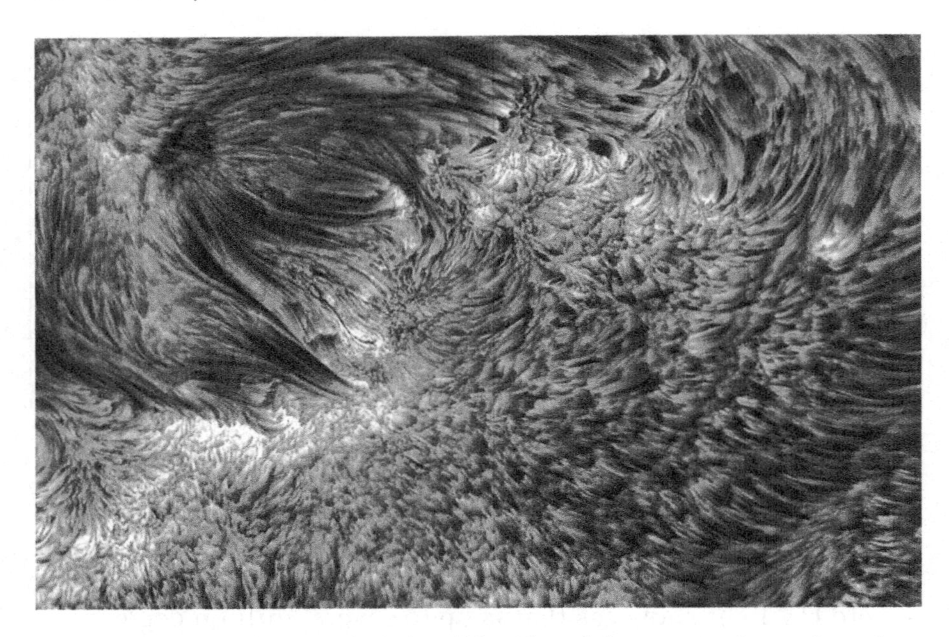

Fig. 4.3 High resolution, slightly off-band H-alpha image of spicules taken on 16th June 2003 with the Swedish 1-m Solar Telescope (*SST*). The spicules are the hairy structures on the *upper left hand corner* of the picture. The dimensions of this image are 65,000 × 45,000 km (Courtesy of SST, Royal Swedish Academy of Science, and LMSAL).

In 2007, Hinode scientists identified two types of spicules – one type moves up and down with periods of 3–5 min, the second type are more jet-like with periods of 10–60 s. The second type, seem to be energised by magnetic reconnection and the scientists thought these may be the cause of rapid heating in the corona.

During 2010, the researchers obtained new observations from the Atmospheric Imaging Assembly (AIA) on NASA's recently launched Solar Dynamics Observatory and NASA's Focal Plane Package for the Solar Optical Telescope (SOT) on the Japanese Hinode satellite to retest their ideas. The high spatial and temporal resolution of the newer instruments was crucial in revealing this previously hidden coronal mass supply.

The scientists have now confirmed that a new class of spicules exists. This new type moves much faster and is shorter-lived than traditional spicules. These spicules shoot upwards at 100 km/s before transferring their energy to the coronal gases. It is now believed that this energy helps maintain the very high temperature of the corona. The observations reveal, for the first time, the one-to-one connection between plasma that is heated to millions of degrees and the spicules that insert this plasma into the corona.

Prominences and Filaments

Prominences are gigantic plumes of dense gas rising up through the chromosphere from the photosphere. They are suspended above the photosphere by magnetic fields and can usually reach heights over 100,000 km with temperatures from 6,000°C to 10,000°C (this is cooler than in the corona). Prominences come in a variety of shapes and sizes; some only last for a few hours while others persist for days. These flame-like structures can be seen projecting out from the limb of the Sun during total solar eclipses; or they can also be seen any time the Sun is shining through hydrogen-alpha solar telescopes. Prominences are the most spectacular events that amateurs can observe. The interesting thing with observing prominences is that they change shape hourly.

The largest prominence on record was observed by the Solar and Heliospheric Observatory (SOHO) in 2010 and was estimated at over 700,000 km long – roughly the radius of the Sun.

Prominences are almost always associated with sunspots and therefore show a variation with the 11-year sunspot cycle. They come in two main types – active and quiescent (Fig. 4.4).

Active prominences

- are energetic and eruptive.
- often extend into the corona or beyond.
- have short lives.
- appear as a surge, spray, jet or loop.

Quiescent prominences

- are dormant.
- are slow to change appearance.
- last longer than active prominences.
- appear like a mound or hill.

Active prominences tend to be narrow at their base, and resemble jagged flames, as they are ejected violently from the chromosphere. They commonly attain heights from 80,000 to 400,000 km, though they have in some cases reached heights of more than 800,000 km – more than one solar radius. Their

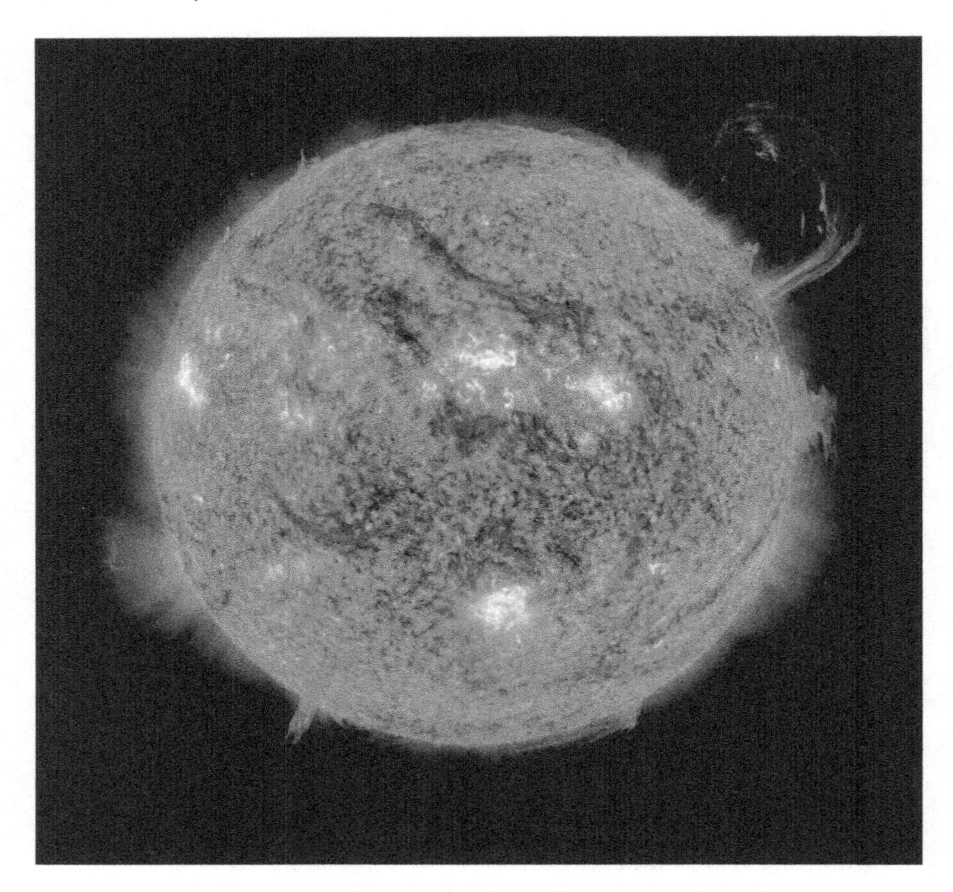

Fig. 4.4 Extreme UV image (at 304 A) of a huge, loop-shaped prominence taken on 14th Sept. 1999 by the SOHO space probe. Prominences are huge clouds of relatively cool dense plasma suspended in the Sun's hot, thin chromosphere and corona. At times, they can erupt, escaping the Sun's atmosphere. Emission in this spectral line shows the upper chromosphere at a temperature of about 60,000°C. Every feature in the image traces magnetic field structure. The hottest areas appear almost *white*, while the *darker red* areas indicate cooler temperatures (Credit: NASA/SOHO).

composition is similar to that of plasma in the chromosphere, often with the addition of metallic vapours.

Quiescent prominences occasionally measure several hundreds of thousands of kilometres across at their base. Their height does not normally exceed some 24,000–48,000 km. Sometimes they appear in the form of enormous clouds suspended in the Sun's atmosphere. They change very slowly, and sometimes survive for several weeks.

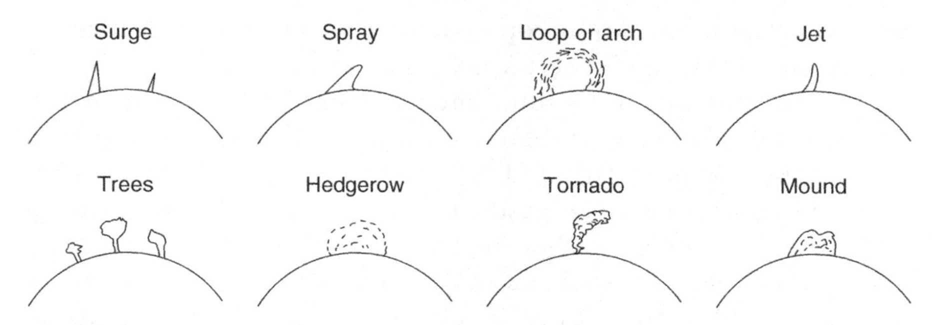

Fig. 4.5 Prominences can be classified according to their shape.

Prominences usually take the shape of the magnetic field that supports them. As the magnetic field changes, so does the shape of the prominence. Like faculae, prominences are not restricted to the zones in which sunspots appear. Prominences can be classified in several ways, but the most useful for amateurs is by their shape. See Fig. 4.5.

All active prominences, such as surges, sprays, loops and jets, display fast structural changes and violent motion. They have life of minutes to hours and are related to sunspots and flares, and show considerable speed.

Surges are straight or slightly curved spikes, which are shot out from a small bright, flare at between 100 and 200 km/s. They can reach heights of more than 200,000 km in the corona before fading out. Many small surges originate near the penumbral borders of sunspots and are directed radially away from the spot along magnetic field lines. It is likely that their energy comes from a burst in the underlying magnetic field.

Sprays are a more eruptive flare, reaching speeds of 500–1200 km/s within a few minutes. They can reach heights of greater than 600,000 km. Spray material can spread out and cover a wide volume of atmosphere. On the solar limb, sprays can appear as an expanding bright mound.

Jets are much thinner than a surge or spray.

The **spectacular loops** are intimately related to flares and are often referred to as post-flare loops. The 'legs' of loops can be traced back to sunspots of opposite polarities. A loop system may last for several hours during which time they expand and reach a height of

50,000 km or more. Arch-loop systems have been seen in extreme ultraviolet (EUV) and in soft x-ray images (Fig. 4.6).

Viewed at the Sun's limb, prominences appear bright red in H-alpha telescopes but when seen against the disc of the Sun, prominences appear dark and are called filaments. Filaments are elongated clouds of cooler gases suspended above the Sun by magnetic forces. They are rather unstable and often break away from the Sun. Filaments associated with a sunspot group are often narrow, dark, and winding lines. They often exist within a longitudinal magnetic neutral line, between two regions with opposite polarity. When you see a filament, you are really looking down on top of a prominence. Filaments are usually found in two latitude belts in the two hemispheres – one in higher latitudes near the polar regions and the other in active mid-latitudes (Figs. 4.7 and 4.8).

Many of the large filaments are subject to violent interruptions of their normal development. A dense filament will frequently vanish, sometimes to reappear with its original shape

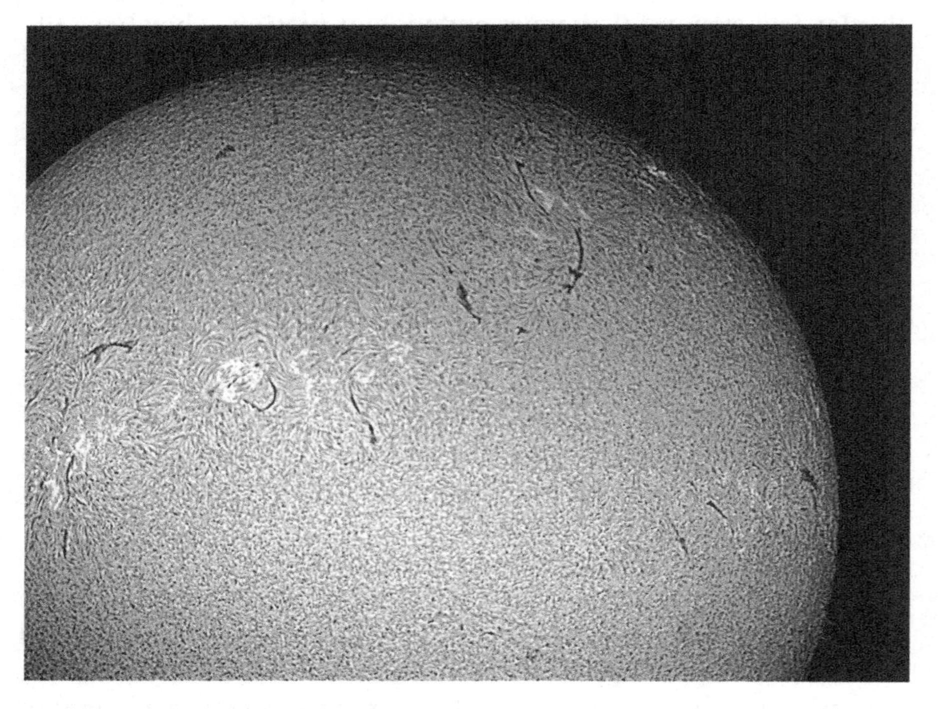

Fig. 4.6 Prominences in the shape of 'trees' on the solar limb. Notice the dark filament on the *left side* of the disc (Credit: Stephen Ramsden).

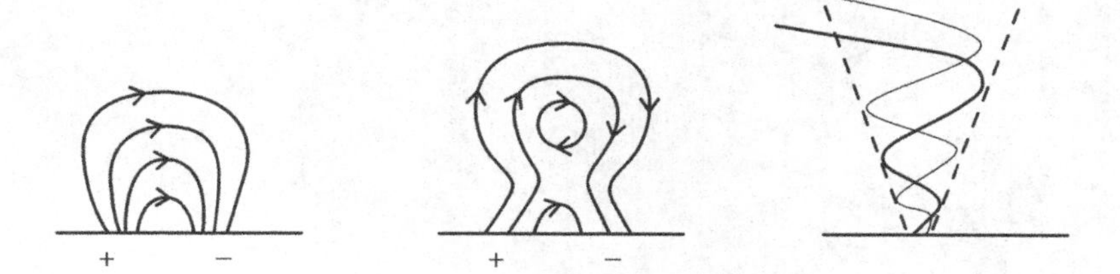

Fig. 4.7 Magnetic field lines associated with some prominences.

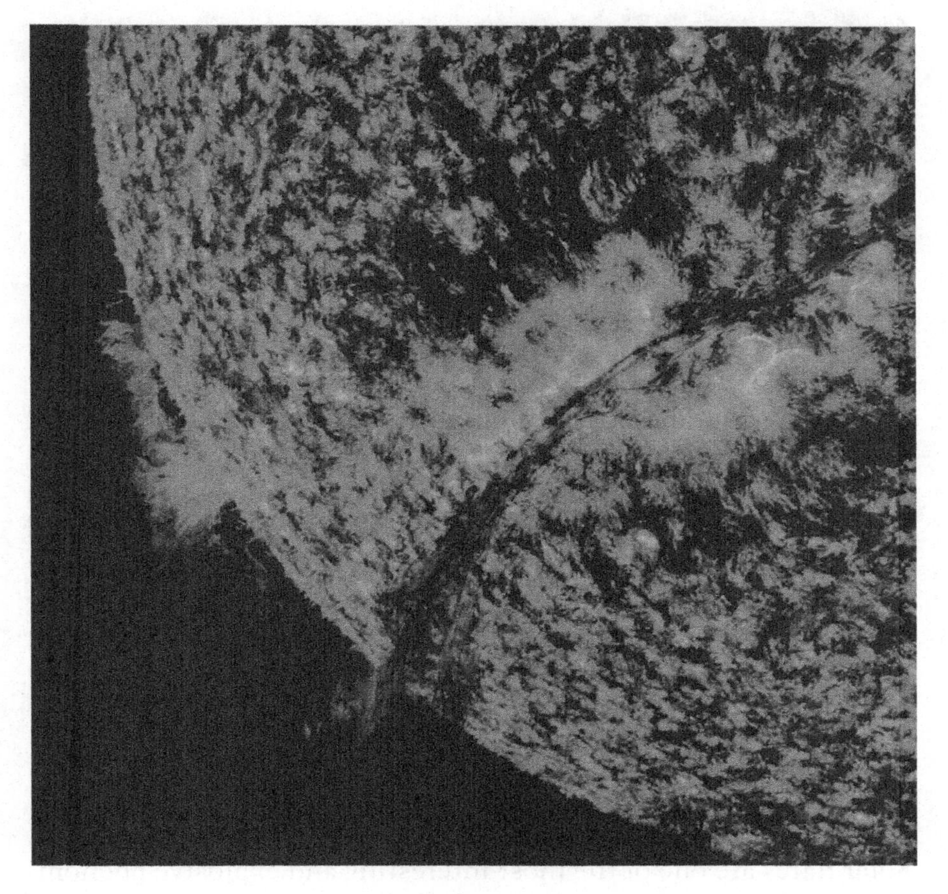

Fig. 4.8 A magnetic filament snaking around the Sun's SE limb stretching more than 700,000 km (a full solar radius). NASA's Solar Dynamics Observatory took the above picture during the early hours of 6th December 2010 in EUV light. Long filaments like this one have been known to collapse with explosive results when they hit the solar surface below (Credit: NASA/SDO).

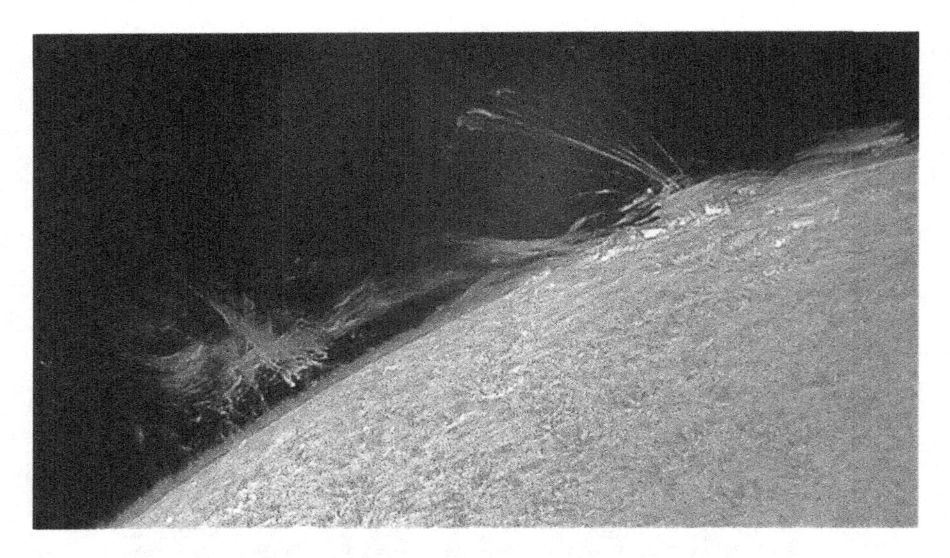

Fig. 4.9 Prominences along limb of the Sun (Credit: NASA).

and density after a lapse of many days. Sometimes violent internal motions disturb the filament and it explodes. The explosion lifts its material high above the chromosphere in the space of a few hours.

After a prominence eruption occurs, the ejected material begins to cool down and condense, flowing back onto the underlying surface. Sometimes the material flow is down both sides of a loop, and occasionally it can be seen rising up one side and returning down another. Prominences do not descend uniformly but in 'jerks'. Prominences also show twisting and untwisting motions in their structure due to interactions between magnetic field lines (Fig. 4.9).

Solar Flares

Solar flares are one of the most interesting and explosive phenomena on the Sun. Flares are bright events that occur suddenly, releasing enormous amounts of energy into the Sun's atmosphere. The first ever solar flare observed was by Richard Carrington and Richard Hodgson in September 1859 who independently saw an intense, bright white light emerging from within a sunspot group.

We know now that flares commonly occur on the Sun. They emit radiation virtually across the entire electromagnetic spectrum, from radio waves at the long wavelength end, through optical emission to x-rays and gamma rays at the short wavelength end. The energy released can be as much as 10^{25} J each second. This energy is ten million times greater than the energy released from a volcanic explosion on Earth. On the other hand, it is less than one tenth of the total energy emitted by the Sun every second. The energy heats the solar plasma to tens of millions of degrees and accelerates electrons, protons and heavy ions to near the speed of light.

The frequency of solar flares varies from several per day when the Sun is 'active' to less than one per week when the Sun is 'quiet'. Large flares are less frequent than small ones. The number of flares seems to vary with the sunspot cycle – basically, more sunspots means more flares.

Solar flares affect all layers of the Sun's atmosphere (photosphere, chromosphere and corona). In the photosphere, only the very energetic are visible in white light (flares are in fact difficult to see against the bright emission from the photosphere). In the chromosphere, there are enormous emissions and structural changes in EUV and x-ray wavelengths. In the corona, flare emissions are strong in the x-ray wavelengths particularly around active regions. In the radio region, flares are marked by various types of emissions from all layers of the atmosphere.

Specialized instruments are used to detect the radiations emitted during a flare. The radio and optical emissions from flares can be observed with telescopes on the Earth. Energetic emissions such as x-rays and gamma rays require telescopes located on spacecraft, since these emissions cannot penetrate Earth's atmosphere (Fig. 4.10).

Cause of Flares

There are a number of competing theories about the cause of flares (it is an area of ongoing research). Flares all occur in regions of strong magnetic fields, and it is the magnetic energy that fuels them. It is believed that a flare occurs when the magnetic field in an active region becomes 'stressed' and magnetic energy builds-up

Fig. 4.10 Close-up view of a loop type solar flare as seen by the TRACE spacecraft in September 2005 (Credit: NASA/TRACE).

pressure as it becomes stored. Instability occurs within the built-up region and an explosion occurs high into the atmosphere. Initially, this release is seen as a sudden brightening or flash in a small active region. Within a minute of the initial flash there is a very sudden increase in brightness followed by a rapid expansion of the flaring region. In the case of very energetic flares there are also large releases of x-ray and microwave radiation. Some flares show a pre-heating phase just prior to the main flare.

For those wanting more detail: Flares are often located between regions of opposite magnetic polarity, and separated by magnetic neutral lines. Magnetic fields pointing in opposite directions interact and merge with each other in a process called 'magnetic reconnection'. Reconnection involves a realignment of magnetic fields, where an area of one magnetic polarity breaks earlier links, and connects with the nearest region of opposite polarity. On the Sun, this often happens when the north pole of the new dipole emerges close to the south pole of an old dipole. In this process the magnetic lines of force experience a strong tension force that snaps them apart with a sudden release of energy. The release of the free magnetic energy provides both thermal and non-thermal energy to the surrounding plasma, accelerating the charged particles. We see this energy release in the form of a plage brightening or solar flare. See Fig. 4.11.

Most confined flares don't last long and cool by conduction or radiation into the cooler chromosphere. The more eruptive flares emit radiation for several hours and often form large magnetic loops that may break apart or reconnect gain. Magnetic reconnection occurs quickly at lower levels and more slowly at higher levels. Sometimes a pair of bright loops may form.

Solar flares often extend out into the corona - the outermost layer of the Sun's atmosphere. These flares have temperatures

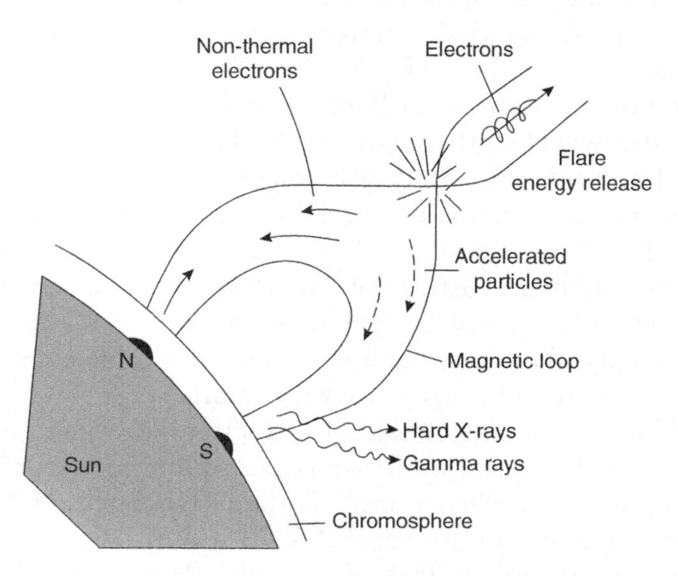

Fig. 4.11 Diagram showing movement and emissions during a solar x-ray flare.

reaching 10–20 million degrees and can be high as 100 million degrees.

The corona is not uniformly bright, but is concentrated around the equator in loop-shaped features. These bright loops are located within and connect areas of strong magnetic field (active regions). Sunspots are located within these active regions and flares occur in them as well.

Classifying Flares

Flares vary in the type of radiation they emit, but they also vary in shape, size, and temperature. Flares are classified according to the intensity of their emissions in optical, radio or x-ray radiation. In the optical region, where flares are usually observed in H-alpha light, classification is by the letter S (sub-flare), and sub-classes F (faint), N (normal) and B (bright). Numerals 1, 2, 3, 4, indicate, increasing order of importance depending on the peak area of the flare.

For example, a 3B flare would cover an area of 12.5–24.7 sq. deg. and be bright (B).

In the radio region, the same letters and numerals are used but instead of area being used, intensity at 5 GHz (6 cm wavelength) in solar flux units (sfu) is used.

Scientists classify solar flares according to their x-ray brightness in the wavelength range 1–8 Å. There are five categories: X-class flares are big; they are major events that can trigger planet-wide radio blackouts and long-lasting radiation storms. M-class flares are medium-sized; they can cause brief radio blackouts that affect Earth's polar regions. Minor radiation storms sometimes follow an M-class flare. Compared to X- and M-class events, C-class flares are small with few noticeable consequences here on Earth. A and B flares are very small.

Soft x-ray flares are classified by a letter followed by a number (the value of the peak flux in watts per square metre or W/m^2 at 100–800 pm x-rays). For example, a flare designated X5.2 is a soft x-ray class flare with intensity 5.2×10^{-4} W/m^2. See Table 4.1.

On 4th November 2003 a large solar flare occurred, initially measuring X28 it was later upgraded to X45. Other large flares of

Table 4.1 Flare classification

Optical (H-alpha) flare		Radio flare		Soft x-ray flare	
Class	Intensity (sq. deg)	Class	Intensity (sfu)	Class	Intensity (W/m^2)
S	2.0	S	5	A	10^{-8}–10^{-7}
1	2.0–5.1	1	10	B	10^{-7}–10^{-6}
2	5.2–12.4	2	300	C	10^{-6}–10^{-5}
3	12.5–24.7	3	3,000	M	10^{-5}–10^{-4}
4	>24.7	4	30,000	X	$>10^{-4}$

note occurred on 2 April 2001 (X20), 28 October 2003 (X17), 7 September 2005 (X17) and 15 February 2011 (X2) (Fig. 4.12).

Did You Know?

A number of space probes have solar flares as their main observation target.

The Yohkoh spacecraft has been observing the Sun with a variety of instruments since its launch in 1991. The observations have spanned a period from one solar maximum to the next. Two instruments are used for monitoring flares – one a soft x-ray telescope and the other a hard x-ray telescope.

The GOES satellites are in geostationary orbits around Earth and have measured soft x-ray emissions from the Sun since the mid 1970s. GOES x-ray observations are commonly used to classify flares with A, B, C, M and X representing different powers of 10.

RHESSI is a spacecraft designed to image solar flares in the soft x-ray band up to gamma rays and to provide high-resolution spectroscopy up to gamma ray energies. X-ray spectra provide a way to distinguish the radiation emitted by hot, thermal electrons from that emitted by accelerated, non-thermal electrons.

The Hinode spacecraft launched by Japan in 2006 is used to observe flares in precise detail. The Hinode observations completed so far indicate that many flares are accompanied by white light emissions. Hinode scientists found connections between white light emissions and hard x-ray emissions. Hard x-rays are emitted when accelerated electrons impact the dense atmosphere near the solar surface. The findings strongly suggest that highly accelerated electrons are responsible for producing white light emissions. Hinodes instruments are supplied by an international group of countries and focus on the powerful magnetic fields thought to be the source of solar flares. Such studies assist scientists to better determine the cause of flares and to predict future flares.

The Atmospheric Imaging Assembly (AIA) on board the SDO spacecraft has recently observed a number of small flares that have generated instabilities and waves with clearly observed effects over the solar surface. The AIA instrument is capable of making full-disc images in eight different temperature bands. This allows scientists to observe entire events that are very difficult to discern by looking in a single temperature band, at a slower rate, or over a more limited field of view.

On 5th December 2006 a group of solar physicists using data from the Stereo spacecraft orbiting the Sun detected a jet of pure neutral hydrogen atoms emanating from an X-class solar flare. No other elements were present, not even helium. Pure hydrogen streamed past the spacecraft for a full 90 min.

Although spacecraft are monitoring the Sun for flares, there is no certain indication that an active region will produce a flare. However, observations clearly show that many properties of sunspots and active regions are linked with flares. Magnetically complex regions called delta spots seem to produce most large flares.

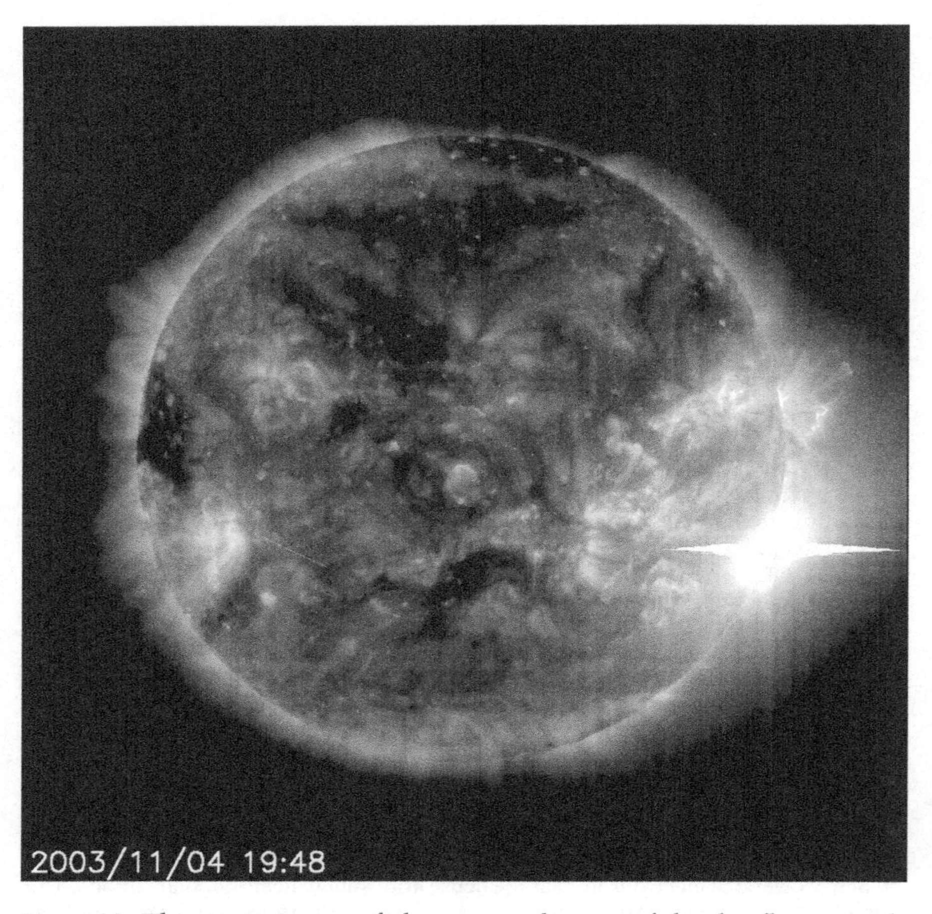

Fig. 4.12 This is an image of the extremely powerful solar flare on 4th November 2003, taken by the SOHO spacecraft in EUV light. The image reveals hot gas in the solar atmosphere in *false colour*. The flare is the bright, white area on the *right edge* of the sun. The *horizontal line* through the flare is the result of the flare's intense light saturating the detector in the EIT instrument (Credit: ESA/NASA).

Hazards of Flares

Solar flares strongly affect the region of space between the Sun and Earth. They produce streams of high-energy particles in the solar wind. These particles get trapped in Earth's magnetosphere and interact with atmospheric particles. The most dangerous emissions from flares are energetic charged particles (primarily high-energy protons) and electromagnetic radiation (primarily x-rays). The solar particles are hazardous to spacecraft and astronauts. They increase the drag on orbiting satellites and produce orbital decay. Energetic particles that get trapped in Earth's magnetic field around the poles cause aurora. The soft x-rays of solar flares increase the ionisation of the upper atmosphere of Earth, which can interfere with GPS systems, short-wave radio communications and heat the outer atmosphere. Hard x-rays from flares can damage the electronics in spacecraft.

Coronal Mass Ejections

In the previous section we saw that a solar flare is defined as a sudden, rapid, and intense variation in brightness. A solar flare occurs when magnetic energy that has built up in the solar atmosphere is suddenly released. Radiation is emitted across virtually the entire electromagnetic spectrum, from radio waves at the long wavelength end, through optical emission to x-rays and gamma rays at the short wavelength end.

As well as flares, the Sun also expels trillions of tons of million-degree hydrogen gas in explosions called **coronal mass ejections** (CMEs). Such clouds are enormous in size and are made up of magnetised plasma gases, consisting primarily of electrons and protons, but may contain small quantities of heavier elements such as helium, oxygen, and even iron. CMEs are rapidly accelerated by magnetised forces to speeds of 20–3,200 km/s with an average speed of 489 km/s (based on LASCO/SOHO measurements taken between 1996 and 2003). Well developed CMEs have extended more than 30 solar radii into space. Such material takes on average about 100 h to reach Earth. The largest CMEs may eject matter with a mass more than 10^{13} kg and kinetic energy about 10^{25} J. The energy of a powerful CME may approach that of a flare, but the material travels slower than a flare because of the increased mass. When such large amounts of solar particles and radiation hit Earth, geomagnetic storms occur.

Coronal mass ejections are usually observed with a coronagraph, such as the LASCO coronagraphs on the SOHO space probe (see Fig. 4.13).

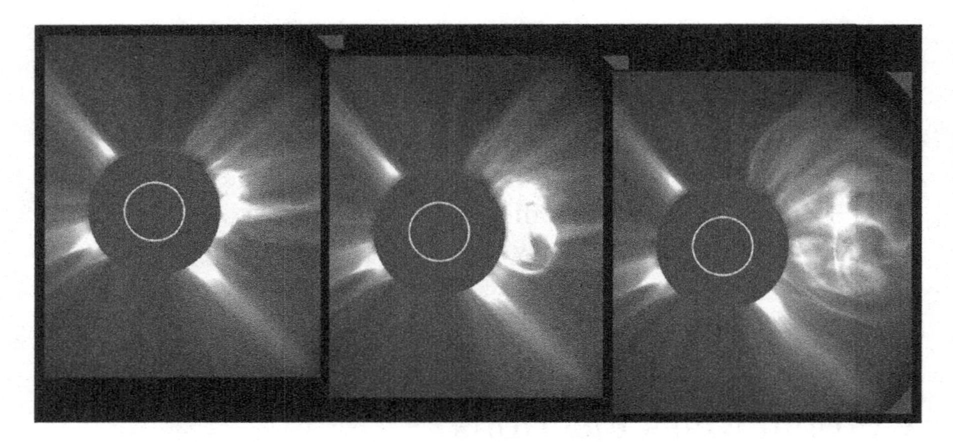

Fig. 4.13 A sequence of LASCO C2 (coronagraph number 2 on SOHO) images showing the evolution of a CME (Coronal Mass Ejection) over a time span of about an hour. The *white circle* indicates the size and position of the Sun (which has been blotted out) (Credit: NASA/SOHO).

Coronal mass ejections are often associated with other forms of solar activity, most notably solar flares and eruptive prominences, but a causal relationship has not been established. Most ejections originate from active regions on Sun's surface, such as groupings of sunspots associated with frequent flares. CMEs occur during both the maxima and minima of solar activity, but they are less frequent during solar minima.

R. Tousey of the Naval Research Laboratory (USA) using the Seventh Orbiting Solar Observatory (OSO-7) made the first detection of a CME on 14th December 1971. Scientists at the Harvard-Smithsonian Centre for Astrophysics saw one of the most recent significant CMEs on 1st August 2010 during solar cycle 24. The scientists observed a series of four large CMEs emanating from the Earth-facing hemisphere of the Sun. The initial event was generated by an eruption associated with sunspot 1092, a sunspot that was large enough to be seen without the aid of a solar telescope. Particles from the ejected material reached Earth on 4th August and a geomagnetic storm resulted in which significant auroras were seen around the poles of Earth.

Most CMEs have been observed using coronagraphs on board space probes. For example, more than 10,000 CMEs have been observed with the Large Angle and Spectrometric Coronagraph (LASCO) aboard the SOHO spacecraft since its launch in 1996.

Coronagraphs C1 and C2 on the twin STEREO spacecraft (launched in 2006) also make regular observations of CMEs. Earth based coronagraphs are limited by brightness and variation of the sky, but they are used to compliment space-based coronagraphs.

Cause of CMEs

Recent scientific research has shown CMEs are associated with enormous changes and disturbances in the coronal magnetic field. The phenomenon of magnetic reconnection or reconfiguration is believed to be responsible for CME and solar flares. **Magnetic reconnection** involves the rearrangement of magnetic field lines when two oppositely directed magnetic fields are brought together. This rearrangement produces a sudden burst of energy stored in the original magnetic fields. Magnetic reconnection commonly occurs in loops that become twisted above active regions that contain groups of sunspots. These regions typically contain magnetic fields that are much stronger than average (Fig. 4.14).

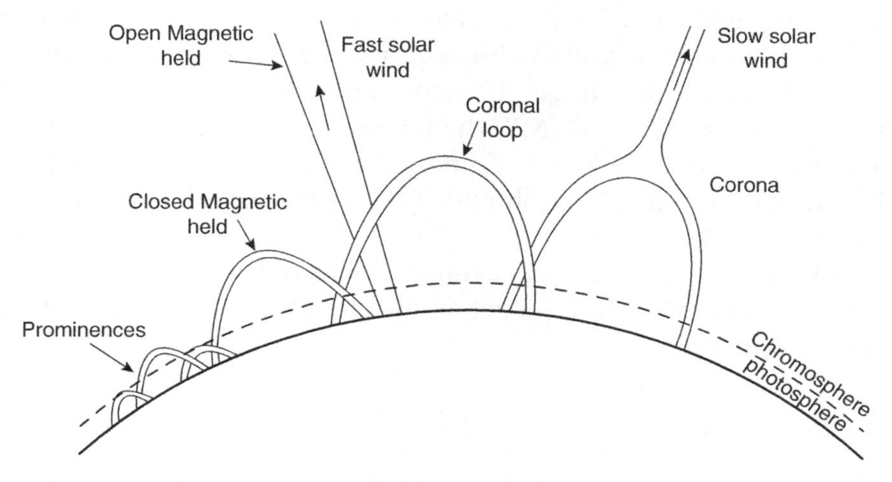

Fig. 4.14 Magnetic structures within the corona. The closed loops are anchored in the photosphere at footprints of opposite magnetic polarity. Such loops can be filled with hot plasma that shines at EUV and x-ray wavelengths. The loops twist, rise, shear, and interact before releasing magnetic energy that can heat the corona and power flares or coronal mass ejections. Some magnetic fields are open and allow the solar wind to burst into interplanetary space.

A CME ejection usually starts with an initial pre-acceleration phase characterized by a slow rising motion, a release of magnetic energy and the emission of soft x-rays. This is followed by a period of rapid acceleration away from the Sun until a constant velocity is reached. Some 'balloon' CMEs, usually the slowest ones, lack this three-stage evolution, instead accelerating slowly and continuously throughout their flight. Even for CMEs with a well-defined acceleration stage, the pre-acceleration stage is often absent, or perhaps unobservable. The duration of these stages can be as short as a few seconds or as long as an hour. In each stage there is a release of soft x-rays and radio waves. This is why CMEs are usually observed using these wavelengths.

A typical coronal mass ejection may have any or all of three distinctive features: a cavity or void of low electron density, a dense core (often a prominence, which appears as a bright region on coronagraph images embedded in this cavity), and a bright leading edge (Figs. 4.15 and 4.16).

The amount of energy produced by a CME is about the same as that produced by a solar flare. However, in a CME the energy is used to lift the expelled mass against the Sun's gravity, whereas in a flare the energy is used to accelerate particles that subsequently release intense x-rays and radio radiation that travel into interplanetary space. Because huge pieces of corona are ripped away from the Sun as a result of a CME, the intensity of its radiation at EUV and x-ray wavelengths is temporarily reduced after the CME. Instruments on the Yohkoh and SOHO spacecraft have observed this dimming.

Research by scientists using Yohkoh data has linked some CMEs to x-ray blow outs – such blow outs do not show any visible chromospheric or photospheric activity.

Impact on Earth

Material from CMEs is often directed towards Earth. It takes on average about 100 h after the eruption for this material to reach Earth. During this time CMEs interact with the solar wind and the interplanetary magnetic field. Fast CMEs travel at about 500 km/s and drive a shock wave through space. The shock wave of the travelling mass of solar energetic particles causes a geomagnetic

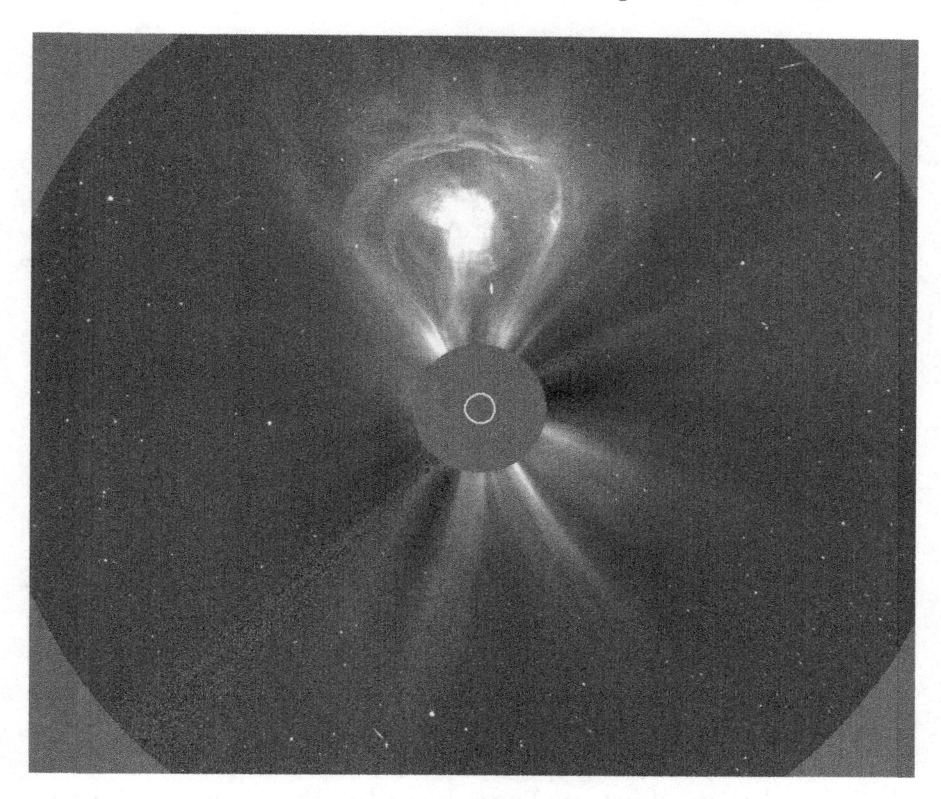

Fig. 4.15 This 'light bulb' shaped Coronal Mass Ejection shows the three classical parts of a CME: leading edge, void, and core. The image was taken on 27th February 2000 by the LASCO C3 coronagraph. Notice that the corona is not uniformly bright, but is concentrated around the solar equator in loop-shaped features (Credit: NASA/SOHO).

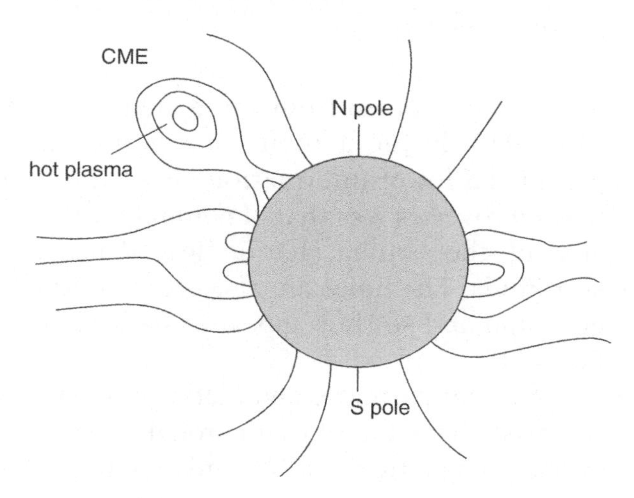

Fig. 4.16 Magnetic field lines around the Sun with a Coronal Mass Ejection.

storm that may disrupt the Earth's magnetosphere, compressing it on the dayside and extending the night-side magnetic tail. The geomagnetic storm can cause particularly strong auroras in large regions around Earth's magnetic poles. These are also known as the Northern Lights (Aurora Borealis) in the northern hemisphere, and the Southern Lights (Aurora Australis) in the southern hemisphere (see chapter 8). Coronal mass ejections, along with solar flares of other origin, can disrupt radio transmissions and cause damage to satellites and electrical transmission line facilities, resulting in potentially massive and long-lasting power blackouts.

On 13th March 1989 a CME knocked out the power to all of Quebec in Canada. Six million people lost electrical power for at least 9 h on a very cold night. Auroras were reported as far south as Texas in the USA. The cost of the strike exceeded tens of millions of dollars.

Humans travelling in space or in aeroplanes are exposed to increased radiation levels when a CME sends radiation towards Earth. Short-term damage might include skin irritation. Long-term consequences might include an increased risk of developing skin cancer.

In recent years, there has been increased interest in studying CMEs and their effect on Earth linked with studies on climate change. Remember, not all CMEs are directed towards Earth (Fig. 4.17).

Coronal Holes

Early images taken from the rocket-borne x-ray telescope on 7th July 1970 during a total solar eclipse and those from Skylab, provided a lot of information about the solar corona. One of the first discoveries was that of coronal holes. A coronal hole is a large area of the corona that is less dense and is cooler than its surrounds. The holes are also areas of very low or even zero x-ray emission and so they appear as dark regions in x-ray images (see Fig. 4.18).

By comparing x-ray pictures with surface magnetograms, scientists have found that coronal holes correspond to a uni-polar or open magnetic field. Not only are the magnetic field lines open,

Fig. 4.17 The clip of the large X2 flare (15th Feb. 2011) seen by Solar Dynamics Observatory (*SDO*) in extreme ultraviolet light has been enlarged and superimposed on SOHO's C2 coronagraph for the same period. This was the largest flare in over 4 years. The coronagraph shows the faint edge of a 'halo' coronal mass ejection (*CME*) as it races away from the Sun and was heading towards Earth (Credit: NASA/SDO).

but they also diverge rapidly and allow the solar wind to flow through into interplanetary space (see Fig. 4.14). Skylab and OSO-7 data confirmed coronal holes as the source of the solar wind in the 1970s.

Coronal holes appear at any time of the solar cycle but they are most common during the declining phase of the cycle. They change shape and size throughout a solar cycle. During solar minimum, coronal holes are mainly found at the Sun's polar regions,

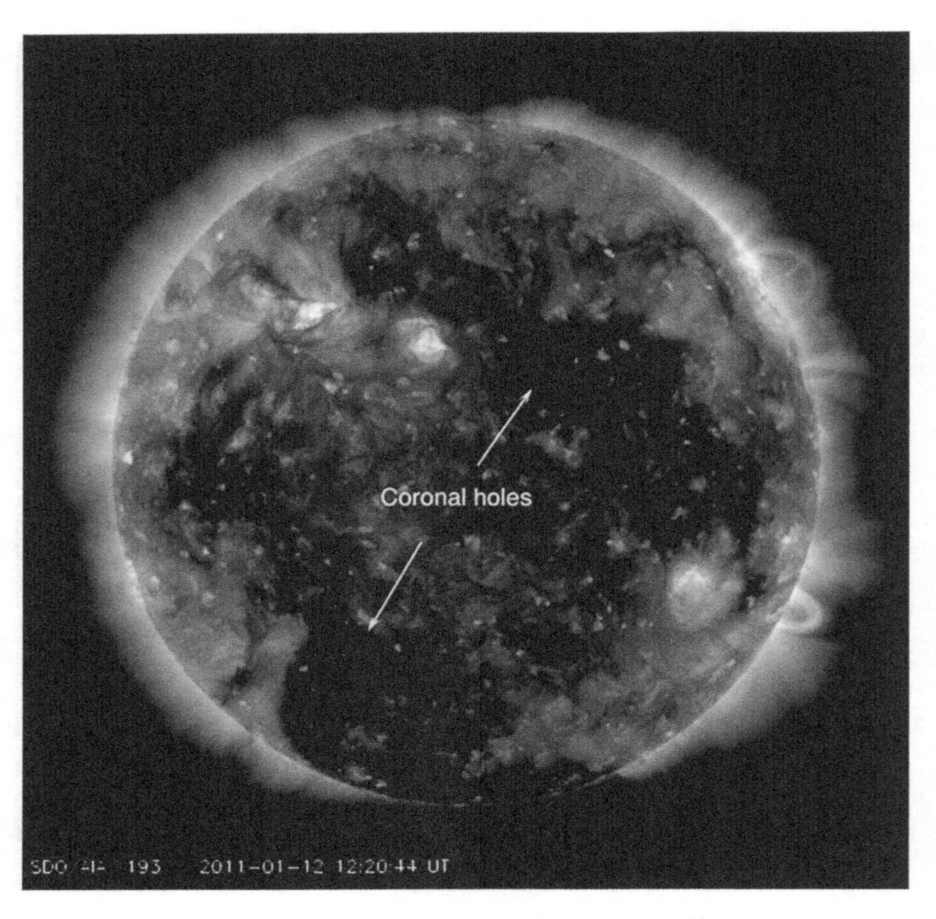

Fig. 4.18 Two coronal holes that developed over several days stand out in this extreme ultraviolet (*EUV*) image taken of the Sun from SDO's AIA instrument on 12th Jan. 2010. Coronal holes are areas of the Sun's surface that are the source of open magnetic field lines that are projected way out into space. They are also the source regions of the fast solar wind, which is characterised by a relatively steady speed of approximately 800 km/s (Credit: NASA/SDO).

but they can be located anywhere on the Sun during solar maximum.

The duration of coronal holes vary from a few hours to several rotations of the Sun depending on the size of the hole. Near the end of their life, coronal holes are seen as bright x-ray emissions (Fig. 4.19).

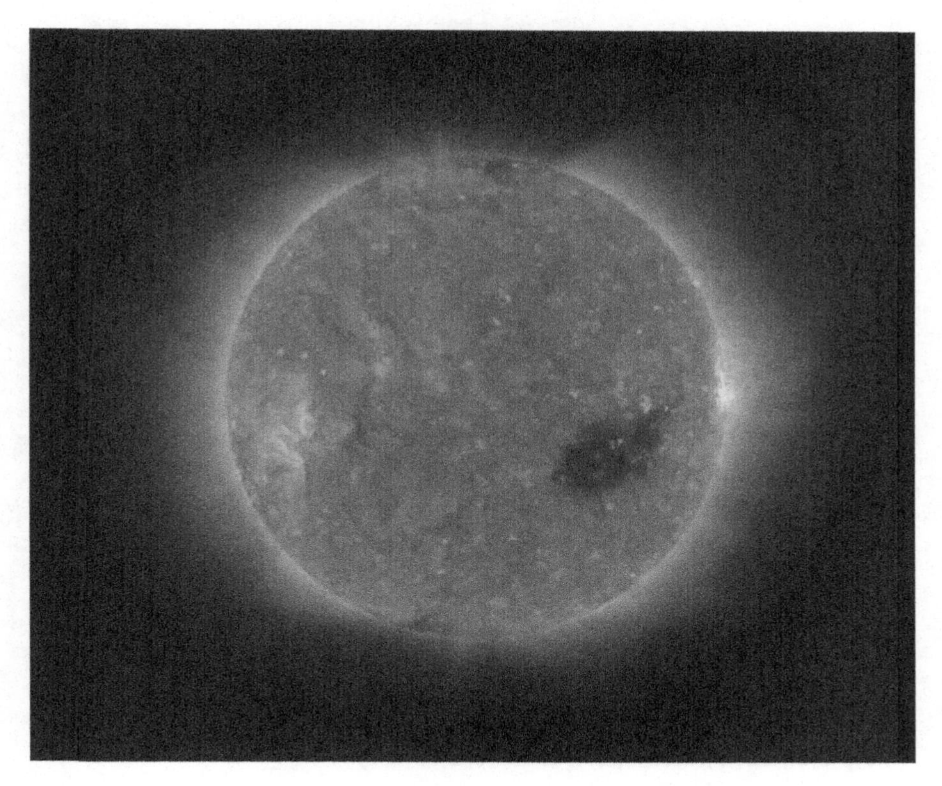

Fig. 4.19 A coronal hole as seen by the Stereo spacecraft in UV light on 25th May 2007. The hole is the dark feature *right of centre* (Credit: NASA/ Stereo).

Observations of coronal holes by the Skylab and Yohkoh spacecrafts have been used to determine the rotation rate of the outer corona. The data suggest coronal holes display differential rotation, that is, holes at different latitudes rotate at different rates. The LASCO instruments on board SOHO have also been used to determine the rotation rate of the corona at different heights. From 2.5 to 15 solar radii the corona displays a radially rigid rotation of 27.5 days.

Coronal Bright Spots

X-ray images of the Sun often show numerous x-ray bright spots. These are seen over the entire surface, although the number in the equatorial region is greater than nearer the poles. Their size is up to

10,000 km and they last for only a few hours. These bright spots are closely related to the photospheric magnetic field and bipolar regions. They spread out in time and become less bright.

High-resolution images from the XRT instrument on board the Hinode spacecraft, in March 2007, have helped scientists see that these x-ray-bright points are actually 'loop' structures demonstrating intense activity.

Web Notes

For info about the Dynamics of solar spicules go to: http://www.its.caltech.edu/~citsolar/paper1/Guadeloupe.html

For info about solar flares go to: http://solarscience.msfc.nasa.gov/flares.shtml. For info and images on CME try: http://www.nasa.gov/stereo/

For info on coronal holes see: http://helios.gsfc.nasa.gov/chole.html

There are a lot of images/pictures/videos available for viewing on the web if you do a search on solar flares, spicules, and coronal mass ejections, and coronal holes.

5. Eclipses and Transits

In this chapter we shall examine two different types of astronomical phenomena associated with the Sun. Both these phenomena are observable from Earth and are spectacular in their own right. The first is the solar eclipse, and the second, transits of the Sun.

Solar Eclipses

An eclipse occurs when one celestial body passes in front of another, dimming or obscuring its light. There are two main types of eclipses – a solar eclipse and a lunar eclipse. Eclipses can only occur when the Sun, Earth and Moon are all in a straight line. A lunar eclipse occurs when the Moon passes into the shadow cast by the Earth; at this time the Moon is situated directly opposite the Sun from Earth. In this chapter we shall be concerned with a solar eclipse.

A **solar eclipse** occurs when the Moon passes in front of the Sun and the Sun's light is blocked from reaching Earth. A solar eclipse is really an occultation of the Sun by the Moon. Such an event is rare and can only occur near a new Moon, that is, when the Moon is between the Earth and Sun. At this time the Moon is crossing over the ecliptic (the path the Sun takes across the sky) (Fig. 5.2).

As seen from Earth, both the Sun and Moon have approximately the same apparent diameter. As a result, when the Moon passes in front of the Sun it can hide it completely from view. If you happen to be at the right place on the Earth's surface, you experience darkness for a few minutes. The darkness is really the Moon's shadow falling on part of the Earth. The area of darkness is only small and moves slowly across the Earth's surface as the

J. Wilkinson, *New Eyes on the Sun*, Astronomers' Universe,
DOI 10.1007/978-3-642-22839-1_5, © Springer-Verlag Berlin Heidelberg 2012

Fig. 5.1 A solar eclipse occurs when the Moon passes in front of the Sun. The Sun's bright surface is blocked enabling the corona and chromosphere to be seen in all there glory.

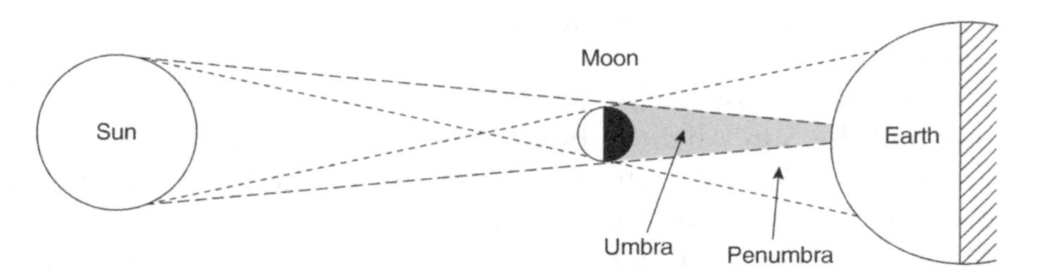

Fig. 5.2 An eclipse of the Sun by the Moon occurs when the Moon passes directly between the Sun and Earth. The Moon's shadow falls on Earth's surface.

Moon moves. Observer's either side of the darkness region experience partial darkness (Fig. 5.2).

A **total eclipse** of the Sun occurs when the Moon completely blocks out the Sun. Totality can last from a few seconds to several minutes. During this time, the Moon is directly between the Earth and Sun. The central part of the Moon's shadow, the umbra, traces a curved path across Earth's surface. This path is only about 270 km wide. The shadow either side of the umbra is partially dark and is called the penumbra. Observer's in the penumbra region will not see totality, but will see a **partial solar eclipse**, where part of the Sun's surface is still visible. See Figs. 5.3 and 5.4.

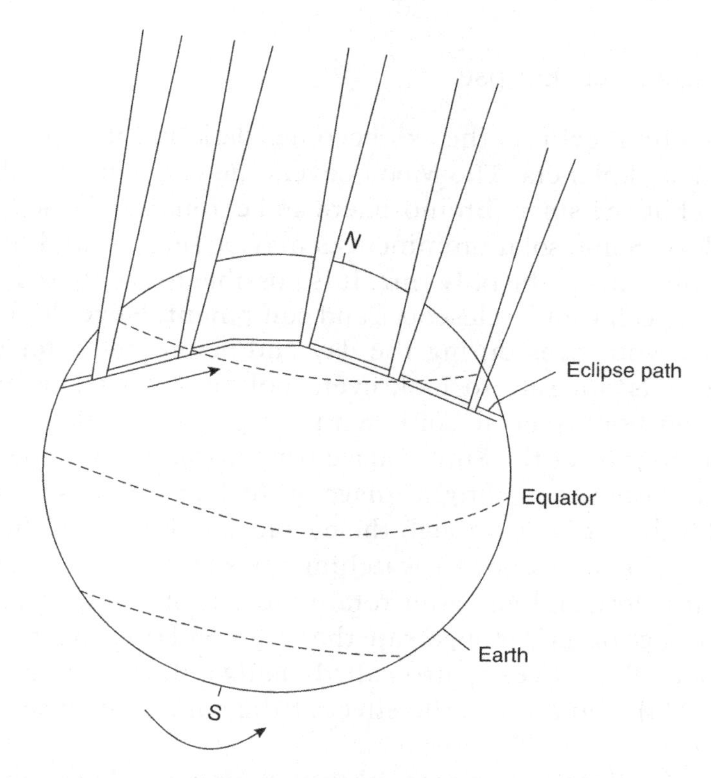

Fig. 5.3 Movement of the Moon's shadow over the spherical Earth.

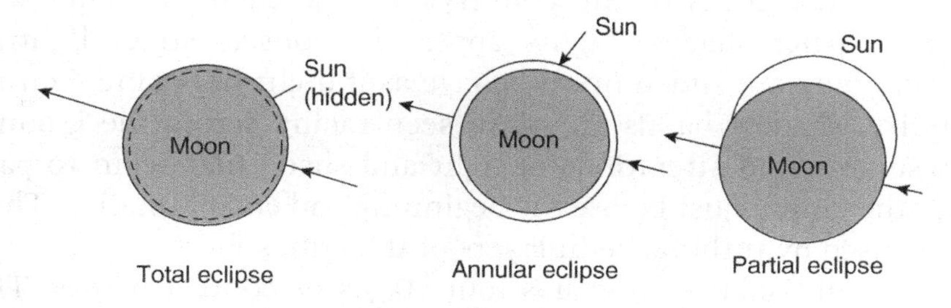

Fig. 5.4 Solar eclipses can be total, annular or partial, depending on how much of the Sun is covered by the Moon.

Annular solar eclipses happen when the distance between the Sun and Moon is such that the Moon does not completely block out the Sun's entire disc. Such an event occurs when the Moon is at apogee (furthest from Earth). During this event a bright ring (or annulus) of the Sun's uneclipsed surface surrounds the black circle of the Moon.

Viewing a Solar Eclipse

During a total eclipse, the sky becomes dark and it is possible to see stars and planets. The Moon covers the entire photosphere of the Sun but the solar chromosphere and corona can be seen in all their glory. Some solar prominences may be seen around the limb of the Sun. This is the only time it is possible to see these features without special solar telescopes and equipment. Never look at the Sun with your eyes during the day and before and after a total eclipse. It is not safe to look even look at a partial or annular eclipse without special equipment. One way to view eclipses safely is to project the Sun's image from a small telescope onto a piece of white card. A bright image of the Sun can be seen on the card. But do not look through the eyepiece to line align the Sun – alignment must be done by watching the shadow on the screen.

Just before and just after totality, beads of sunlight shooting over the edge of the Moon, create the "diamond ring effect" around the Moon. This effect is also called "Baily's Beads" after F. Baily (1774–1844) who noticed the effect at the annular eclipse of 1836 (Fig. 5.5).

Total eclipses give rise to other phenomenon. At the approach of totality one has the impression of nightfall; the temperature drops, and colours of the countryside lose their brilliancy and assume paler shades. Lights appear in houses, street lighting might come on, and animals behave as if night has come. During totality, shadow bands are often seen racing across the ground. These are rapid alterations of light and shade that seem to pass over the ground just before the beginning and end of totality. They are caused by turbulence in layers of the atmosphere.

A total solar eclipse has four stages or contact points. The eclipse begins at first contact, which is the moment when the Moon's disc first touches the solar one. First contact can be

Fig. 5.5 The Diamond ring effect or Baily's Beads around the Sun during and eclipse.

detected only through a telescope prepared for solar viewing. Over the next hour, the partial phase unfolds, as the Moon steadily moves across the Sun's disc. Second contact is the start of totality when the sky becomes dark. Third contact is the end of totality and there is a bright burst of light from behind the lunar limb as daylight begins again. Fourth contact occurs when the last part of the Moon's limb has lost contact with the Sun's disc (Fig. 5.6).

Solar Eclipse Case Study

During 2009, there were two solar eclipses, one annular and the other total. The annular eclipse took place on 26th January. The path of annularity was mainly across water, from the South Atlantic to the Indian Ocean. The path crossed land in Sumatra and Borneo in Indonesia. Greatest eclipse of about 8 min occurred over the Indian Ocean about halfway between Australia and Madagascar. Observers in most parts of Australia saw some degree of partiality

Fig. 5.6 The Moon's shadow on the Earth as taken from the Mir spacecraft in 1999 during a solar eclipse (Credit: NASA).

although the Sun was low to the horizon or below the horizon at maximum eclipse.

Figure 5.7 shows the view at maximum partial eclipse as seen from various locations in Australia. The information is given here to show you how such information is presented in Astronomy Year Books.

The second solar eclipse of 2009 was a total one that occurred over the Pacific Ocean and China on 22 July. The path of totality began on the west coast of India and travelled northeast touching Nepal's eastern border, the northern section of Bangladesh, and most of Bhutan, before crossing over China. The path left land near Shanghai passing below Japan and across the Pacific Ocean. The maximum duration of 6 min and 39 s occurred over the Pacific Ocean, however, land based observers in eastern China got around 5 min of totality. Interestingly there have been only four total eclipses this century that have exceeded 6 min.

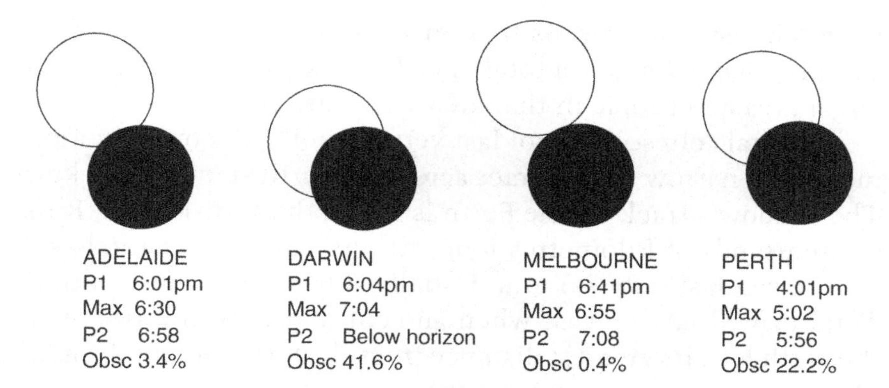

Fig. 5.7 View at maximum partial solar eclipse as seen from various locations in Australia. *P1* local time of first contact, *Max* time of maximum eclipse, *P2* second contact or end of partial eclipse, *Obsc* percentage of solar disc obscured by Moon at maximum eclipse.

Did You Know?
An eclipse of the Sun is one of nature's grandest spectacles. Such events have even stopped wars. In 585 B.C. an eclipse occurred during a battle between the Lydians and Medes of Asian Minor. The frightened soldiers ceased fighting and signed a peace treaty.

Ancient Egyptians believed an eclipse occurred when an underworld serpent swallowed the boat carrying the Sun god Ra, during his daily journey across the sky. The ancient Chinese described an eclipse as a dragon devouring the Sun. Other cultures associated eclipses with outbreaks of disease. A study of records dating back to 763 B.C. from Babylonia and China, showed that many ancient astronomers understood the cause of a solar eclipse.

The oldest records of a solar eclipse probably date back to 3 May 1375 B.C. and 5 March 1123 B.C. The records are uncertain because the information was collected from clay tablets found at the ancient Babylonian city of Ugarit (in modern Syria) and there are disputes about the translation.

Time of a Solar Eclipse

Every month the Moon passes between the Earth and Sun at new Moon phases. So, why doesn't a solar eclipse occur every month? There would be, if the Moon's orbit and the Earth's orbit were in the same plane. But the Moon's orbit is at an angle of about 5° to the Earth's orbit. The Moon is usually above or below the Sun, and there is no eclipse.

Solar eclipses can occur from two to five times a year. Only three of these eclipses in a year can be total. As already mentioned, the umbra shadow on the Earth is never wider than 270 km. It may be much less. This means that only a small part of the world can see any solar eclipse as a total one. The penumbra falls on a much larger area and people in that area see a partial eclipse.

A total eclipse does not last very long. The Moon's revolution makes the narrow shadow race across the earth at over 1,600 km/h. The shadow's track on the Earth is called the eclipse path. It may be thousands of kilometres long. At any one time, a totals solar eclipse can last only 7.5 min. Usually it lasts only a few minutes. Totality will last longest when an eclipse occurs at a time when the earth is at its greatest distance from the Sun (aphelion), and the Moon is at perigee (nearest Earth).

Solar eclipse events seem rare because 300 years can pass before one is seen in the same location again (Fig. 5.8).

Recording a Solar Eclipse

Eclipses can be recorded photographically, either on film or digitally. They can even be video taped. The eclipsed Sun will be small in the frame, but a telephoto lens can enlarge the image. Some observer's take pictures through their telescope. The image of the Sun projected onto a white screen can also be photographed. With any photo, the correct exposure time depends on the film speed and f-ratio. Exposures generally are short rather than long. Prior to totality, photos will need to be taken through special solar filters. The filter can be removed at totality when the Moon blocks the glare from the Sun's disc. Digital and video cameras can make use of the zooming capabilities of their cameras. The best results are obtained with the camera mounted firmly onto a tripod or telescope.

If using a camcorder, make sure the auto-focus function is set to OFF and the focus is manually set to infinity. Use the manual shutter speed to adjust for optimal exposures.

The advantage is using digital cameras is the ability to see what is happening in the viewfinder or LCD monitor as it happens. If you are inexperienced at taking sure photos, it is best to take a number of pictures using different camera settings. Poor photos can be easily deleted after the event. The key to successful eclipse photography is to be well prepared beforehand. If inexperienced, you should seek the help of an experienced person who is a

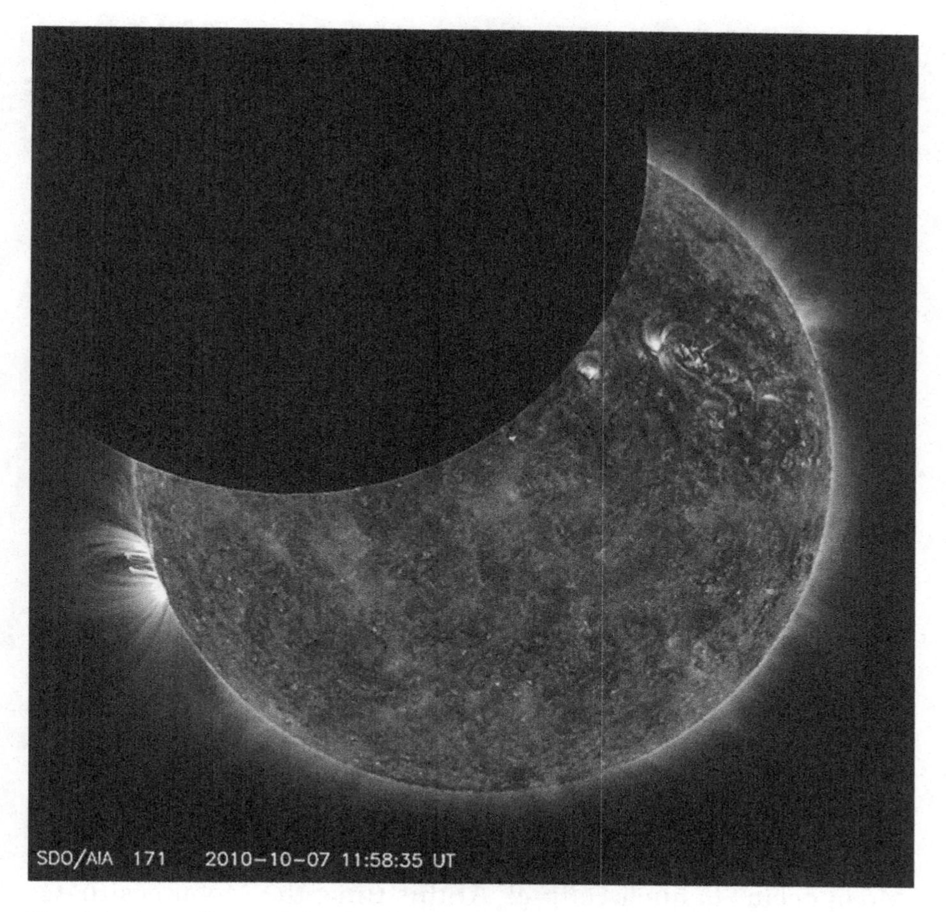

Fig. 5.8 An eclipse of the Sun by the Moon as seen from space by the AIA instrument on board the SDO spacecraft 7th October 2010 (Credit: NASA/SDO).

member of an astronomy club. There are plenty of books on astrophotography that cover photographing eclipses. There are also plenty of solar eclipse images and videos on the internet (Table 5.1).

Lunar Eclipses

Another type of eclipse is the lunar eclipse – it is important to understand how it is different to a solar eclipse.

Table 5.1 Future solar eclipses up to 2015

Date	Type	Location
25 November 2011	Partial	S. Africa, Antarctica, Tasmania, New Zealand
20 May 2012	Annular	Asia, Pacific, N. America
13 November 2012	Total	Australia, NZ, Sthn Pacific, S. America
10 May 2013	Annular	Australia, NZ, C. Pacific
3 November 2013	Hybrid	E. America, S. Europe, Africa
29 April 2014	Annular	S. Indian, Australia, Antarctica
23 October 2014	Partial	N. Pacific, N. America
20 March 2015	Total	Iceland, Europe, N. Africa, N. Asia
13 September 2015	Partial	S. Africa, S. India, Antarctica

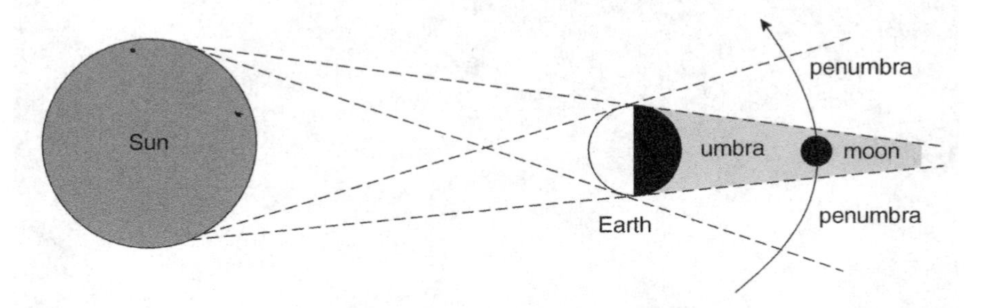

Fig. 5.9 A total eclipse of the Moon occurs when the Moon is completely covered by the central zone (umbra) of the Earth's shadow. This figure is included to show how such an eclipse is different to a solar eclipse.

The Earth casts a cone shaped shadow into space as it revolves around the Sun. This shadow reaches nearly 1,400,000 km behind Earth. When the Moon passes into this shadow an eclipse of the Moon occurs (a lunar eclipse). At this time, the Moon is situated directly on the opposite side of the Earth to the Sun. The shadow has two parts. One part is the umbra or total shadow; the other part is the penumbra, or partial shadow. If the Moon is fully in the umbra, the eclipse is said to be total. If the Moon is only partially in the umbra, the eclipse is partial (Fig 5.9).

For further details on lunar eclipses see the book "The Moon in Close-up" by the author of this book and published by Springer.

Solar Transits

A solar transit is the passage of a planet or other object across the disk of the Sun. Such transits may be thought of as a special kind of eclipse. As seen from Earth, only transits of the inner planets

Mercury and Venus are visible. Transits by planets are far more rare than eclipses of the Sun by the Moon. On the average, there are 13 transits of Mercury each century. In comparison, transits of Venus usually occur in pairs with 8 years separating the two events. However, more than a century elapses between each transit pair.

The first transit ever observed was of the planet Mercury in 1631 by the French astronomer Gassendi. A transit of Venus occurred just 1 month later but Gassendi's attempt to observe it failed because the transit was not visible from Europe. In 1639, Jerimiah Horrocks and William Crabtree became the first to witness a transit of Venus.

In recent years spacecraft like the International Space Station and the space shuttle, have been observed from Earth passing in front of the Sun. Communication satellites regularly pass in front of the Sun and when they do the signals from them get interrupted by the huge microwave radiation emitted by the Sun. In the Northern Hemisphere, solar transits of satellites usually occur in early March and October. In the Southern Hemisphere, solar transits of satellites usually occur in early September and April. The time of day varies mainly with the longitude of the satellite and receiving station, while the exact days vary mainly with the station's latitude. Stations along the equator will experience solar transit at the equinoxes, as that is where geostationary satellites are located directly overhead. Pictures and times of such events are available for viewing on the internet.

Transits by Planets

It is an exciting event to observe the transit of a planet such as Mercury and Venus across the face of the Sun. Transits of Mercury usually occur within several days of 8th May and 10th November each year. At these times Mercury's orbit is inclined 7° to Earth's, and it intersects the ecliptic at two points or nodes which cross the Sun each year on those dates. If Mercury passes through inferior conjunction at that time, a transit will occur. During a May transit, Mercury is closer to Earth and its disc appears at its largest. However, during November transits, Mercury is closer to the Sun and its disc is smaller. Transits of Mercury are expected on 9th May 2016, 11th November 2019, 13th November 2032,

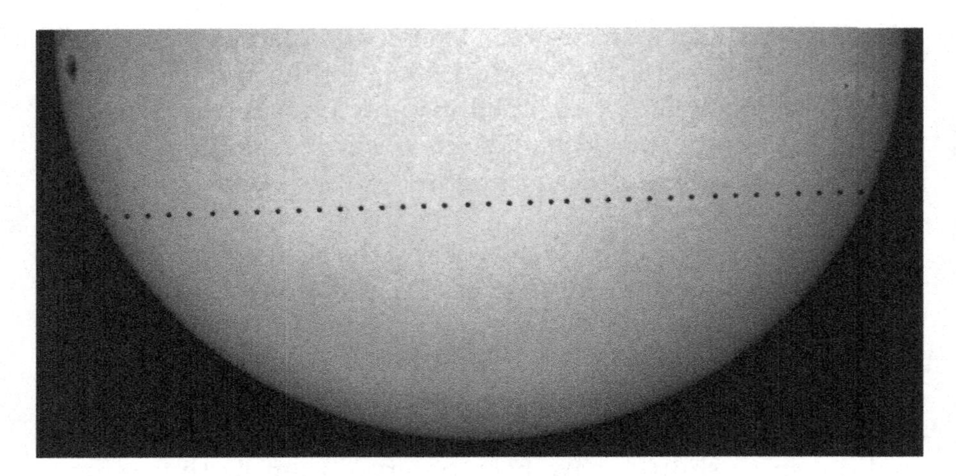

Fig. 5.10 A sequence of images showing the 8th November 2006 transit of Mercury as seen by the SOHO spacecraft. Mercury is about 200 times smaller than the diameter of the Sun as seen from the Earth; hence it appears small in this image (Credit: NASA/SOHO).

7th November 2039, and 7th May 2049. Dates and the best locations to view transits are published in Astronomy Year books (Fig. 5.10).

Transits of the Sun by Venus are more rare than those by Mercury, because the orbit of Venus is much larger than that of Mercury. Only seven transits by Venus have been seen since the invention of the telescope (1631, 1639, 1761, 1769, 1874, 1882, and 2004). A transit occurs when Venus reaches conjunction with the Sun at or near one of its nodes, the longitude where Venus passes through the Earth's orbital plane or ecliptic. Transits by Venus are only possible during early December and June when Venus's orbital nodes pass across the Sun. See Fig. 5.11.

Sequences of transits occur in a pattern that repeats every 243 years, with transits occurring 8 years apart followed by a gap of 121.5 years, then a gap of 8 years and then another long gap of 105.5 years.

The 2004 transit by Venus provided a good deal of interest as scientists attempted to measure the pattern of light dimming as Venus blocked out some of the Sun's light, in order to refine techniques that they hope to use in searching for planets around other stars. Extremely precise measurements were made and it was found that the Sun's light dropped by a mere 0.001 magnitudes.

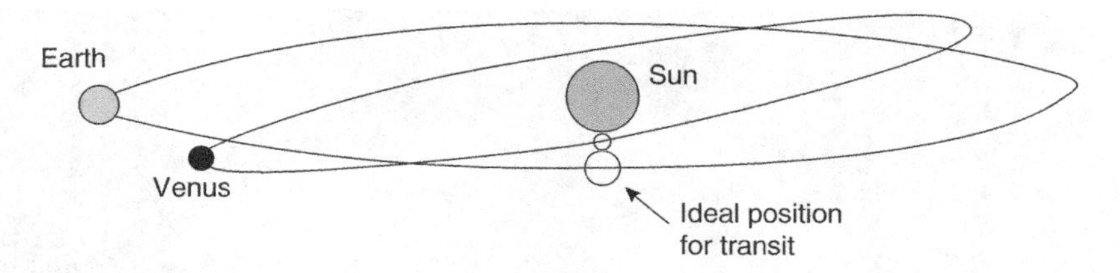

Fig. 5.11 The orbit of Venus is inclined at 3.4° to Earth's. Because of this Venus usually passes above or below the Earth's orbital plane and no transit is visible. The ideal position for a transit is when the two bodies are at conjunction (i.e. at least distance of separation and on same side of Sun).

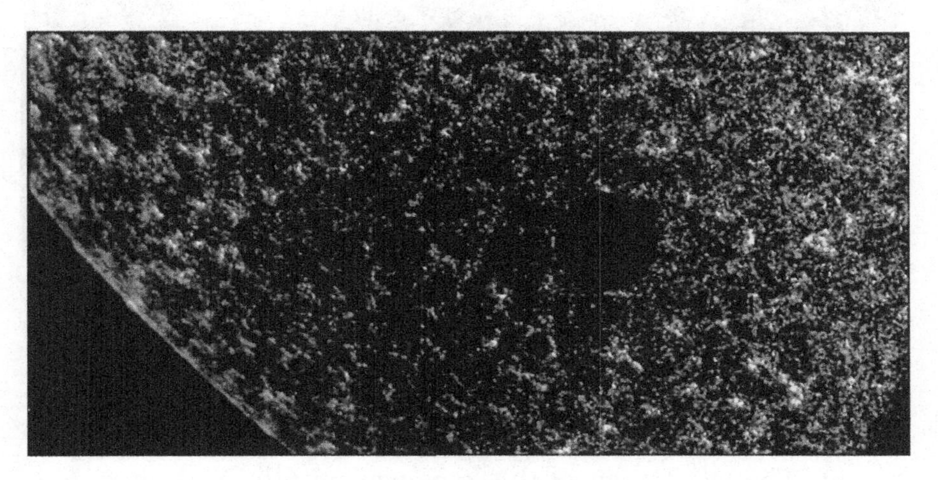

Fig. 5.12 The 8th June 2004 solar transit by Venus. The TRACE satellite captured this image of Venus crossing the face of the Sun as seen from Earth orbit (Credit: NASA/TRACE).

The 2004 transit by Venus was visible from Europe, Africa and Asia. The final stages of the event were also visible from the eastern USA and Canada. Transits of Venus are expected on 6th June 2012, 11th December 2117, 8th December 2125 and 11th June 2247. The 2012 transit will be visible from North America, the Pacific, Asia, Australia, Eastern Europe, and eastern Africa. Details can be found on the internet under "2004 and 2012 Transits of Venus" (Figs 5.12 and 5.13).

Sometimes Venus only grazes the Sun during a transit. At these times it is possible that in some areas of the Earth a full transit can be seen while in other regions there is only a partial

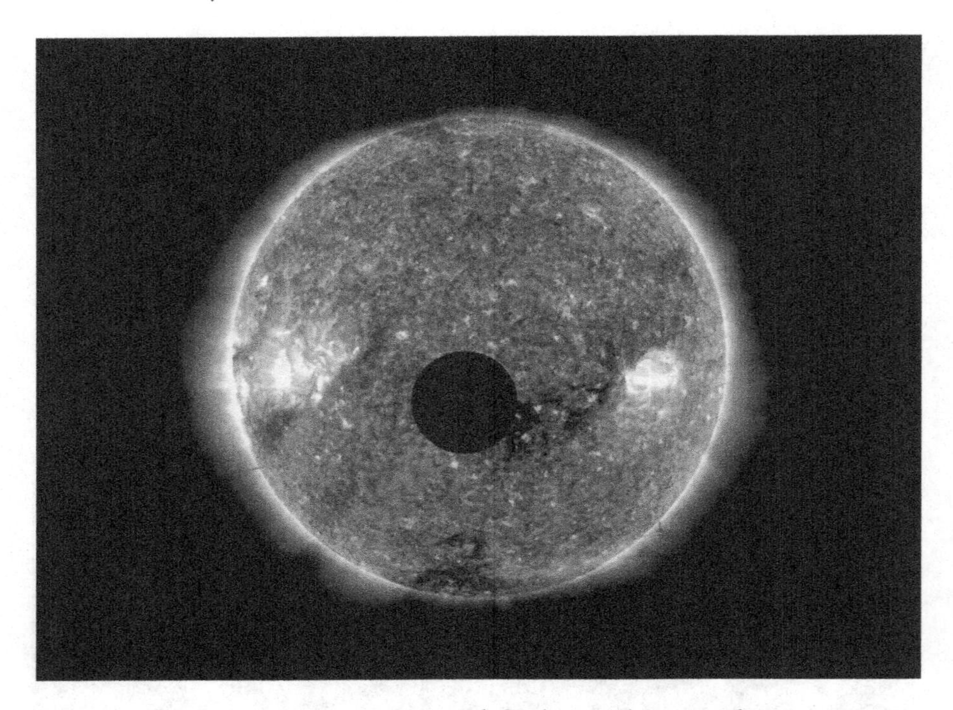

Fig. 5.13 NASA's STEREO spacecraft had gone far enough into space to capture the Moon's transit of the Sun on 25th Feb. 2007. The spacecraft captured the Moon as a dark disc racing across the Sun's face. When the Moon eclipses the Sun as seen from Earth, it's usually described as an eclipse rather than a transit (Credit: NASA/STEREO).

transit. The last transit of this type was on 6th December 1631, and the next such transit will occur on 13th December 2611. It is also possible that a transit of Venus can be seen in some parts of the world as a partial transit, while in others Venus completely misses the Sun.

Observing Transits

The safest way to observe a transit is to project the image of the Sun through a telescope, binoculars, or pinhole onto a white screen. Care must be taken not to look directly at the Sun with your eye, as damage will result. See Chap. 6 for projection and other techniques.

There are four contacts or moments during a transit: when the disc of a planet touches the circumference of the Sun at a single point:

1. First contact: The planet is entirely outside the disc of the Sun, and moving inward.
2. Second contact: The planet is entirely inside the disk of the Sun, and moving further inward.
3. Third contact: The planet is entirely inside the disk of the Sun, and moving outward.
4. Fourth contact: The planet is entirely outside the disk of the Sun, and moving outward.

A fifth named point is that of greatest transit, when the planet is at the middle of its path across the solar disc. It is worthwhile recording the times of each contact to the nearest second. Your location (obtained from GPS signals) needs to be recorded as well. Photographs can be also be taken of these contacts.

Did You Know?

There was a transit of Venus in 1761 and 1769. These transits were used to try to determine the precise value of the astronomical unit (distance from the Sun to Earth) using the method of parallax. James Gregory in "Optica Promota" first described this method in 1663. In an attempt to observe the first transit of the pair, scientists and explorers from Britain, Austria and France travelled to destinations around the world, including Siberia, Norway, Newfoundland and Madagascar. Most of the scientists managed to observe at least part of the transit, but excellent readings were made by Jeremiah Dixon and Charles Mason at the Cape of Good Hope.

Mikhail Lomonosov used his observation of the 1761 transit of Venus from the Petersburg Observatory, to predict the existence of an atmosphere on Venus. Lomonosov detected the refraction of solar rays while observing the transit and inferred that only refraction through an atmosphere could explain the appearance of a light ring around the part of Venus that had not yet come into contact with the Sun's disk during an early stage of the transit.

For the 1769 transit, scientists travelled to Hudson Bay, Baja California, and Norway. Observations were also made from Tahiti by Captain James Cook at a location known as "Point Venus". The Czech astronomer Christian Mayer was invited by Catherine the Great to observe the transit in Saint Petersburg with Anders Johan Lexell, while other members of Russian Academy of Sciences went to other eight locations in Russian Empire.

In 1771, the French astronomer Jerome Lalande, using the combined 1761 and 1769 transit data, calculated the astronomical unit to have a value of 153 million km (±1 million km). This is close to the current accepted value of 150 million km.

Web Notes

For info and movies of solar eclipses: http://umbra.nascom.nasa.gov/eclipse/images/eclipse_images.html
NASA's eclipse site (also useful for transit information): http://eclipse.gsfc.nasa.gov/eclipse.html
For the 2012 solar eclipse see: http://www.solareclipse2012.com/
For videos of past solar eclipses and a world map of future eclipses see: http://www.exploratorium.edu/eclipse/

6. Observing the Sun

This chapter is about methods used by amateur astronomers to view the Sun safely. There are three regions of light that will be discussed – the first involves observing the Sun in white light, the second uses Hydrogen alpha light, while the third uses Calcium K light. This chapter also provides a review of some of the most common types of telescopes used by amateurs to observe the Sun safely as well as observing tips.

Safe Solar Observing

When you are observing the Sun, your first concern should always be eye safety. Serious eye damage can result from even a brief glimpse of our nearest star. One of the key rules is to never view the Sun directly with the naked eye or with any unfiltered optical device, such as binoculars or a telescope. Do NOT use smoked glass, welding glass, dark sunglasses, blackened film, or polarizing filters.

The famous astronomer Galileo Galilei (1564–1642), looked at the Sun through a telescope 400 years ago and suffered permanent eye damage. If it happened to Galileo, it can happen to you! So please be careful.

Having given a safety warning, there are safe ways to observe the Sun. This chapter will take you through the methods and instruments available for amateur astronomers to use. Observing the Sun can be terrific fun – astronomy in daylight hours. One of the good things about solar observing is that the Sun's surface is constantly changing – it does not appear static like many other astronomical objects.

To observe the Sun you really need an astronomical grade telescope that is not too large. Successful observing with a

J. Wilkinson, *New Eyes on the Sun*, Astronomers' Universe,
DOI 10.1007/978-3-642-22839-1_6, © Springer-Verlag Berlin Heidelberg 2012

telescope depends on a number of other factors. One key factor is good seeing conditions. The atmosphere of Earth is often turbulent and this causes images of the Sun to look blurred or show rippling heat waves around the limb. The quality of images in telescopes are affected by:

1. Atmospheric conditions.
2. Scattered sunlight as it passes through the telescopes optical system.
3. Temperature differences between the ground and the air, and in the air itself.
4. A poor telescope mounting system causing unwanted vibrations.
5. Air inside the telescope heating up causing internal turbulence.
6. The time of day (early-mid morning or late afternoon are good times).
7. The optical quality of the observing instrument being used.

With a good quality telescope and good seeing conditions, solar observing should be rewarding.

When making solar observations one should keep accurate records of things like, optical equipment (type of telescope, eyepiece, filters, etc.), seeing conditions, date and time of day, and location. Good records are also important when taking photographs of the Sun.

There are different methods used to observe the Sun. These methods will now be discussed in detail.

White Light Solar Observing

The Sun as it appears to the naked eye is said to be in "white light". White light contains all the colours of the visible spectrum (red through to violet) mixed together. When we see the Sun in white light, we are looking at the photosphere of the Sun. Some white light features are of low contrast, that is, there is only a slight difference in brightness between the features and the solar surface. However, white light viewing is interesting and there is much to see and learn about.

Method 1: Solar Projection

Telescopes for observing the Sun usually have an objective lens or mirror less than 100 mm in diameter. The Sun is so big that high magnifications are not required but good resolution is.

One safe way to observe sunspots or eclipses is to project an image of the Sun through a telescope or binoculars onto a white screen or any other plain surface. If you are using a telescope, be sure that any small finder telescope is capped otherwise you might melt any crosshairs inside the finder or be tempted to look through it. If you are using binoculars, keep the lens cover on one of the two tubes. Never look through a telescope or binoculars to aim them at the Sun – partial or total blindness will almost surely result. You should use the instrument's shadow and image of the Sun cast on the screen or ground to align the instrument (your back should face the Sun while doing this). If your telescope has a diagonal, point it to the side. On the screen you should see the disc of the Sun as a bright circle of light. Adjust the distance between the screen and the telescope until the disc is about the size of a small plate. The image will probably be blurred at first; focus your telescope and move the screen until the circle becomes sharp. The larger the image, the less bright it will be. Using this method you can see considerable detail in and around sunspot groups but you will not see any prominences. You may even see some faculae (bright patches). This method is called eyepiece projection (see Fig. 6.1) and it involves a bit of trial and error at first.

The best telescope for eyepiece projection is a refracting telescope (about 60–90 mm diameter) with magnification about 40×. A Newtonian telescope with open tube can be used providing it is not too large (up to 100 mm diameter objective mirror). Never use a catadioptric telescope (i.e. Schmidt-Cassegrain or Maksutov-Cassegrain) with this method because of the risk of heat build-up damaging internal components. If using binoculars, use those that have front lenses about 50 mm in diameter. Binoculars usually are described by a pair of numbers separated by an "×", such as "7 × 30" or "7 × 50"; the first number gives the magnification while the number to the right of the "×" is the diameter of the front lenses in millimetres. Big lenses gather a lot of light, and the heat generated by direct sunlight inside large binoculars can

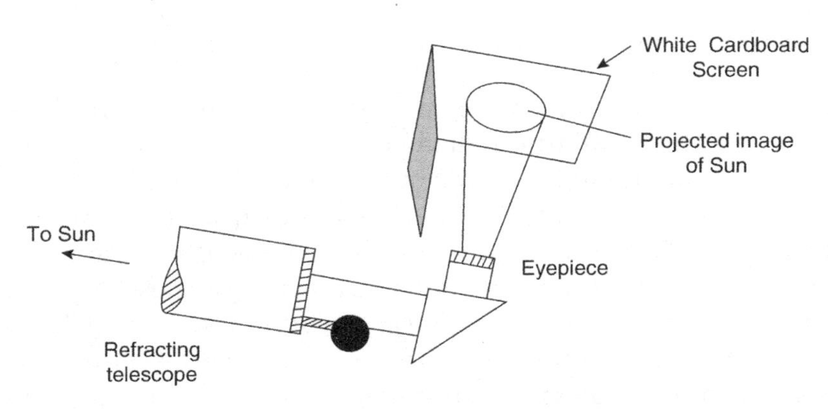

Fig. 6.1 Using eyepiece projection to view sunspots in white light. Never insert a hand, eye, or piece of clothing between the eyepiece and the screen – you might get burnt; and never look through the telescope with your eyes.

damage their optics. It is best to cover one of the lenses with a cap, and use only one lens.

An advantage of using the projection method to view the Sun is that a group of people can view the image at the same time – this is also useful for viewing an eclipse of the Sun. Use a white screen that is very smooth. A shade around the screen would also help improve viewing. Care must be taken when selecting an eyepiece for solar projection – make sure there is no optical cement (glue) holding the eyepiece lenses together – the intense heat could melt the glue. Cheaper eyepieces such as Huygenian and Ramsden eyepieces with a metal casing are good.

You can tell the difference between sunspots and dust on the lens by lightly tapping the telescope to see what moves with the scope and what stays in place on the image.

The image seen can be traced onto white paper to record the position and shape of sunspots. If possible use a logbook. Observations can be done over a period of several days to monitor the daily progress of sunspots. On the other hand you can prepare a number of circles about 90 mm in diameter in your logbook and draw in the position of sunspots. Make your observations at the same time each day as the Sun changes orientation slightly as it moves across the sky. Remember the Sun rotates, as seen from above its north pole, in a counter-clockwise direction, east to west. Sunspots also move from east to west across the solar disc over several days. The direction of

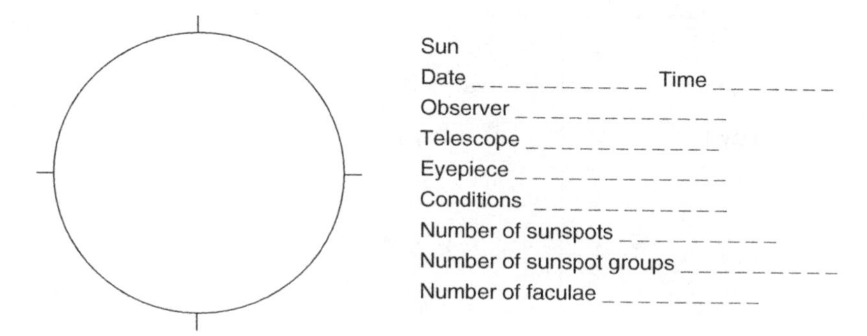

Fig. 6.2 Sample chart to record daily sunspot/faculae positions.

motion of the sunspots therefore tells you which direction on the Sun's image is west (see section on directions at the end of this chapter) (Fig. 6.2).

Daily drawings of sunspot groups can be used to monitor their evolution (change in shape and appearance) as they move across the Sun's disc. Remember a single spot is also a group. Groups can also be classified according to the system in Chap. 3 (Table 3.1) or any other classification system. Take care to record differences between penumbra and the umbra of sunspots. Also note the location of any bright patches or faculae. A red/orange filter in the eyepiece is sometimes useful when counting sunspots, while a blue filter is often useful for observing faculae and granulation. Astronomical societies have sections for solar observing – it is best to join one of these and work in with other amateurs (you also learn a lot from experienced people) (Fig. 6.3).

Method 2: Solar Objective Filters

There are specially made filters that can be used to view the Sun in white light via a telescope. These fit over the FRONT of a telescopes objective lens and reduce the intensity of the sunlight before it enters the tube. It is not safe to use filters that mount in the eyepiece. Objective "white light" solar filters come in two main types. The most durable is a filter made of glass coated with a thin layer of metal (making the filter look like a mirror). The second type of solar filter is a thin, metal-coated foil or mylar film coated with a metal layer. The metal coating is designed to reflect most of the Sun's heat and light away from the telescope.

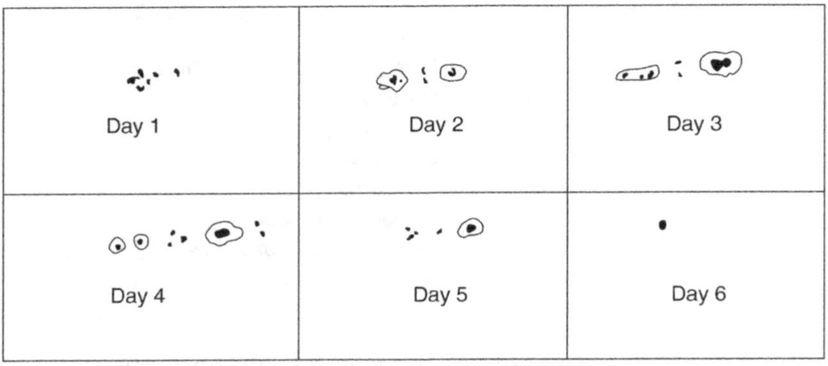

Day 1 Day 2 Day 3

Day 4 Day 5 Day 6

Fig. 6.3 Drawing showing the evolution of a sunspot group over a number of days.

Both types of filters must be mounted in some sort of frame or cell that fits over the front of the telescope's objective. The fit must be secure enough so it does not fall off during use (screws and tape can be used). A word of caution – not any filter will do – you must buy the correct filter from an astronomical supplier. The following are recommended brands: JMB solar filter (Identiview class A glass type), Thousand Oaks (Type 2+ glass), and Baader Astrosolar safety film. Other manufacturers such as Orion and Tuthill supply similar filters. See Fig. 6.4.

The Baader Astrosafety film reduces the Suns intensity by one hundred thousand times. The material making this specialised filter is ion implanted and metallised with a tough, colour-neutral layer on both sides of the film. This ion implantation/metallisation process produces a high-contrast film that stands up to considerable abuse. The coating cannot easily be rubbed off.

The foil or thin film types of filters might look loose – don't pull them tight as stretching will spoil the optical quality and may also open up microscopic cracks in the metal coatings, letting in stray light that will degrade contrast. The film should remain relaxed and a little billowy.

All the above solar filters show the Sun in white light; that is they reduce the amount of sunlight across the entire spectrum. Such filters are suitable to view sunspots, faculae and eclipses but not prominences. The majority of solar filters offer sharp, contrasty views during times of good atmospheric stability, but there

Fig. 6.4 A glass type solar filter mounted on the front of the author's SCT telescope objective lens. Notice that the cap of the finder scope is in place, so the Sun's image does not melt the cross hairs inside and so that you do not look through it. Align the scope using the Sun's shadow of the scope.

are differences between the various types. The JMB glass filter shows the Sun in more neutral shades of yellow-white. The Baader Astrosolar safety film shows the Sun in white with a slightly purplish tinge; while the Thousand Oaks glass filter produces a yellow-orange Sun. Some mylar films produce a bluish or greenish image. The colour produced does not really matter.

Prices of solar filters vary depending on the size required for your telescope, but allow about $80–200. In some cases solar foil or film can be purchased as A4 sheets from astronomical suppliers – it is then up to the purchaser to cut the required shape and make a suitable and safe mounting frame or cell. Glass solar filters are more durable and likely to last longer than film or foil types.

A high-end glass solar filter is a well-made optical flat, deposited with a metal coating. Both sides of the filter must be parallel for best results. Baader Astrosafety film has a high reputation for producing good images.

Whatever filter type you may use, make sure it is stored safely (not able to be scratched) in a dry container. You can regularly inspect the filter for scratches or pinholes by holding it up to the light prior to use. Mylar type filters are best cleaned with a soft brush or compressed air only as the coatings are particularly fragile. A glass filter may be gently cleaned by brushing away dust or dirt with a soft cotton wool ball, blower brush, or lens cleaning solution (such as isopropyl alcohol) and a soft cloth or cotton ball. Do NOT clean the back surface of a glass filter. Since the coating is on the backside of the glass facing the telescope it will rarely need cleaning if handled correctly and kept sealed when not in use. Baader Astrosafety film can be cleaned with a mild detergent and distilled water.

Method 3: The Herschel Wedge

The Herschel wedge is a device that looks like a star diagonal, and fits into the rear of a refracting telescope in the same place as a star diagonal. The eyepiece is inserted into the Herschel wedge just like it is in a star diagonal. So what's the difference? Well, a Herschel wedge contains a thin wedge or prism shaped piece of glass that reflects only about 5% of the incoming sunlight to the eyepiece. A small portion of the heat and light is absorbed by the wedge, but 90–95% is transmitted through the wedge and out its rear. This means the rear of the device gets hot, and caution must be taken to prevent eyes, fingers, etc., from being placed near these rays. Most units have an effective light trap system to dissipate heat and light, which prevents accidental burning or injury. There is usually a secondary filter inside the device to further reduce the sunlight and minimise unwanted reflections. Telescopes using such a device should be pointed away from the Sun and allowed to cool down once observing is finished (Fig. 6.5).

A word of CAUTION – Herschel wedges are only suitable for use with refracting telescopes. Do NOT use them with a Catadioptric (SCT or Mak-Schmidt) telescope – as these scopes will be damaged by internal heat.

Herschel wedges are suitable for viewing the Sun in white light – they produce a sharp and contrasty image, with a black background.

Recommended commercially available Herschel wedges are, the Lunt Solar wedge and the Baader Planetarium Cool-ceramic

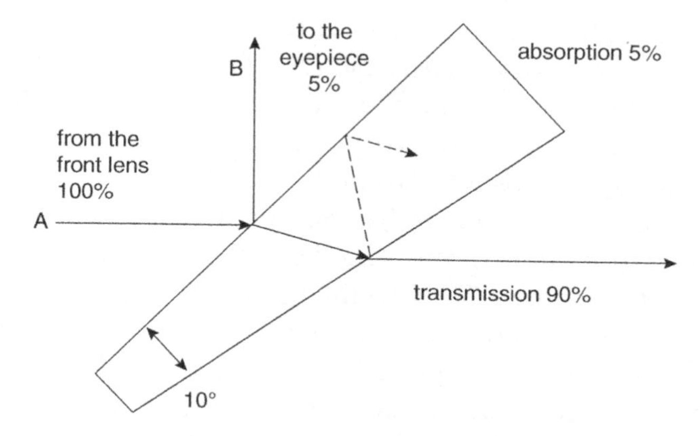

Fig. 6.5 Optical structure of a Herschel wedge.

Fig. 6.6 A refracting telescope with a Lunt Herschel wedge attached (in place of the star diagonal). The eyepiece fits into the wedge (Courtesy of Stephen Ramsden).

Herschel Prism. Both devices are designed for a 2 in. eyepiece holder. The ideal scope to fit them to would be a quality refractor of 80–100 mm aperture.

The back of the Lunt solar wedge has a red circular heat sink that is located just under a series of vent slits for heat dissipation (see Fig. 6.6). There is no way of catching the full light from the Sun through any part of this device. The wedge/prism is fully enclosed in the diagonal and there is no reflection of a disc or any other full brightness image to be seen. Price is about $300 but accessories might boost that figure.

The Baader Herschel wedge incorporates an innovative internal light trap and ceramic tile backplate with a multi-layer perforated steel screen. The bright light and heat are harmlessly diffused. Another unique aspect of the new cool-ceramic safety Herschel wedge is the integrated solar finder. The Sun's defocused image can be readily seen projected onto the translucent ceramic backplate, permitting fast and easy aiming. There are two versions of the wedge – one for visual use and the other for photographic use; and there are different support filters available depending on your desired use. The device is ONLY designed for use with refracting telescopes. This is probably the safest wedge on the market, and it is suitable to view sunspots, faculae, and granules on the Sun's surface. It provides the best white light views of the Sun out of all the devices so far discussed in this chapter. Price is around $500–600.

Did You Know?

Do you know who invented the Herschel wedge? The name that first comes to mind is Sir William Herschel (1738–1822). He was a professional musician and amateur astronomer of German origin. He came to England as young man in 1766. He had an amateur's interest in mathematics and astronomy, both of which he studied passionately. He was too poor to afford a telescope, so he made his own using a bronze mirror. He made other telescopes, one of which was used to discover what he thought was a comet, but which turned out to be a new planet – Uranus. Herschel was elected a Fellow of the Royal Society and George III appointed him court astronomer. Herschel lived near Windsor where he devoted himself to astronomical observation, assisted by his sister Caroline Herschel (1750–1848). He was the first to make measurements of the relative position of the two components of double stars. Herschel engaged in research on clusters and nebulae, of which barely a hundred were known before his time; and he increased the number by more than 2,500. He also investigated the distribution of stars in all directions of the sky and drew up a profile of the galaxy showing the enormous concentration in the median plane.

Sir William Herschel was an active solar observer. He discovered infrared radiation and attempted to view the sun through darkened and smoked glass but many cracked. But Sir William Herschel did NOT invent the Herschel wedge.

The concept of the solar wedge was first conceived by Sir William's son, John Frederick William Herschel (1792–1871), in a book published in 1847, 25 years after his father's death. His original idea was not as an accessory but as part of a primary optical system (involving an unsilvered mirror) to disperse excess energy as heat.

John was a noted astronomer too, and continued his father's work, particularly in the field of double stars. The wedge is further described in John Herschel's 1861 book "The Telescope". John also devised another solar eyepiece where the optical path was blocked by a rapidly rotating metal disc cut with narrow radial slots, thus passing a fraction of the light. For further information about the history of solar devices, see "Web Notes" at the end of this chapter.

Hydrogen Alpha Solar Observing

H-alpha is a specific red visible spectral line created by hydrogen with a wavelength of 6,563 A (656.3 nm). The line forms as a result of emission when a hydrogen electron falls from its third to second lowest energy level. The significance of the H-alpha line, as far as the Sun goes, is that it is one of the brighter emission lines of the chromosphere. The reddish-pink colour of the chromosphere is due to the dominance of the H-alpha line and its location in the red part of the electromagnetic spectrum. Solar features visible in H-alpha light include the prominences, filaments, spicules, plage, flares and sunspots.

It is possible to view the Sun in H-alpha light using a telescope fitted with a H-alpha filter. A H-alpha filter is an optical filter designed to transmit a narrow bandwidth of light centred on the H-alpha wavelength. These filters are manufactured by coating several layers of a vacuum-deposited material onto a substrate such as optical glass. The layers are selected to produce interference effects that filter out all wavelengths except the required band. Alternatively, a Fabry-Perot etalon may be used as the narrow band filter in conjunction with a blocking filter (BF) or energy rejection filter (ERF) to pass only a narrow (<1.0 A) range of wavelengths of light centred round the H-alpha emission line. The physics of the etalon and the interference filters are similar, relying on the constructive and destructive interference of light reflecting between two surfaces. Energy rejection filters remove unwanted infrared and ultraviolet light to further reduce the bandpass of the telescope.

Due to the high velocities sometimes associated with features visible in H-alpha light (such as fast moving prominences and mass ejections), solar H-alpha etalons are able to be tuned (by tilting, changing the temperature or air pressure or spacing of the etalon) to cope with the associated Doppler shift. Doppler shift is the change in wavelength of spectral lines when an object is moving toward or away from the observer.

In the past, the cost of H-alpha filters has been very high ($3,000+) and beyond the reach of amateurs, however in recent years new technology has resulted in reduced prices and now H-alpha filtered telescopes are popular among amateur solar observers (and available for $600–2,000).

What's Visible in H-alpha?

In H-alpha light, the Sun's chromosphere has a red appearance. There is an irregular fringe of red light, running all the way around the edge of the Sun's disc. At high power (especially in the wings of H-alpha), the individual spicules making up this fringe are sometimes visible as tiny narrow jets of light. On the disc, spicules look like small dark mottle with low contrast.

The standout feature seen in H-alpha light is the prominences. These flame-like structures can be seen projecting out from the solar limb – at active times there are a lot of prominences, while at quiet times there are few. Viewed around the Sun's limb, prominences appear bright red in H-alpha, but when seen against the disc of the Sun, prominences appear dark (snake-like) and are called filaments.

Sunspots are visible in H-alpha, but their penumbrae are lower in contrast than in white light. Frequently, fibrils (fine dark lines) will be seen near sunspots, tracing out the nearby magnetic field lines. Also visible at times in or near active regions are bright patchy areas called plage or flocculi. Plage occurs in the chromosphere and stands out brightly in H-alpha. Faculae are similar to plage but occur in the photosphere and are a white-light feature (not H-alpha). Flares sometimes appear as a sudden brightening of existing plage.

H-alpha solar telescopes/filters with a bandpass of just less than 1 A enable the observer to see prominences and some surface detail. The narrower the bandpass, the better the visibility of surface detail (and the higher the cost of the telescope/filter). Ideal bandpass is around 0.7 A. Some scopes have two filters (called double stacking) to achieve a lower bandpass around 0.5 A. But double stacking nearly doubles the cost of the telescope and brightness is reduced.

H-alpha Telescopes

Solar telescopes that are designed to work in H-alpha light are available commercially and are very good. A number of companies such as Coronado, Lunt and Solarscope make dedicated solar telescopes – these usually can't be used for anything else but observing the Sun. In this section we shall review some of these scopes.

The basic design of a H-alpha solar telescope is similar to that of a normal refracting telescope with an achromat objective lens at the front and an eyepiece at the rear. Covering the front lens is an energy rejection filter (ERF). Inside the telescope tube is an etalon or blocking filter (BF) or both. Some different types of designs are shown in Fig. 6.7. Front loading H-alpha scopes have the etalon at the front, while end-loading types have the etalon at the rear before the eyepiece.

Coronado PST

One of the most popular solar telescopes is the Coronado Personal Solar Telescope or PST. It can be brought for around US$500 and can be easily mounted on a solid camera tripod. It is not necessary to mount such a scope on an equatorial mount. The scope is very well made and comes in a solid travel case. Starting with such a scope has the advantages of low cost and portability. It can be easily moved about and ready for use within minutes (Table 6.1).

The PST contains an energy rejection filter built into the coating of the objective. The etalon filter is placed between the eyepiece and objective at the junction of the black box and the main tube. The focusing knob changes the position of a diagonal prism in the black box. There is also a small blocking filter (5 mm) near the bottom of the eyepiece tube (similar to design 1 in Fig. 6.7).

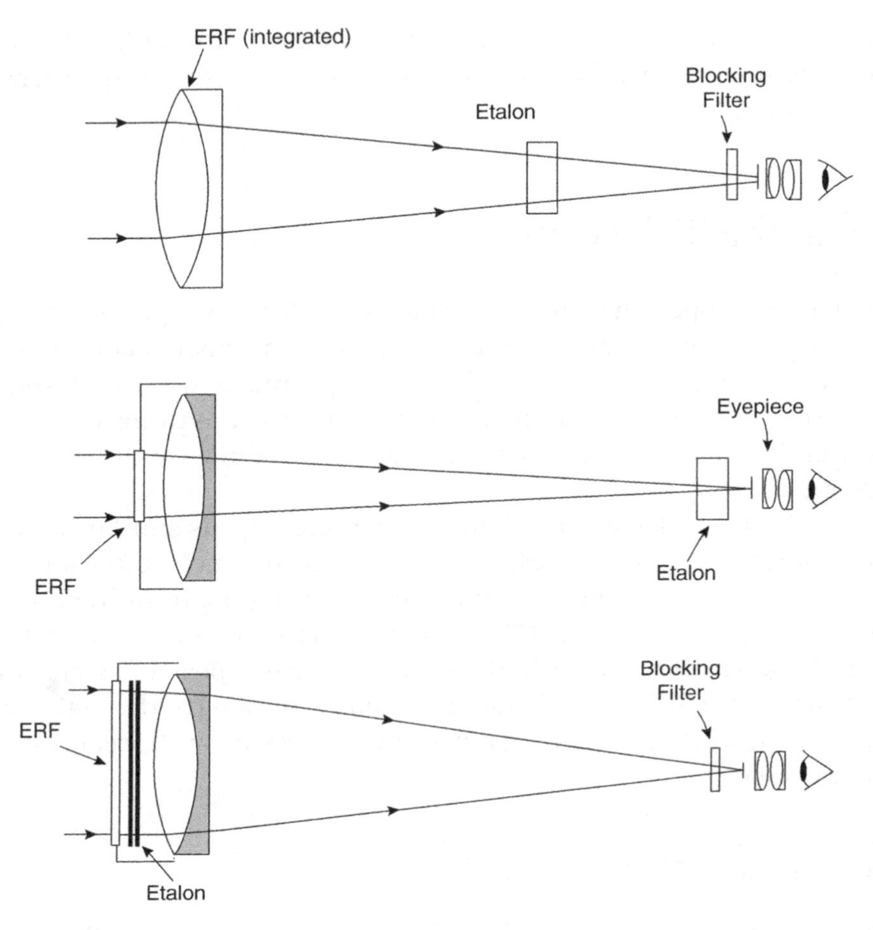

Fig. 6.7 Basic designs of some solar telescopes.

Table 6.1 Specifications of the Coronado PST

Aperture	40 mm
Focal length	400 mm
Focal ratio	f/10
Bandwidth	<1.0 A
Weight	1.5 kg (3 lb)
Standard eyepiece	20 mm
Eyepiece size/magnification	20 mm = 20×, 15 mm = 27×, 12 mm = 33×, 10 mm = 40×

Coronado has built a small and portable alt-azimuth "MALTA" mount for use with the PST. The threads on the bottom of the PST will accept any standard 1/4 in. × 20 threads per inch photographic or astronomy mounting accessories. The author has found an aluminium camera tripod is better than the tabletop "MALTA" mount.

Once mounted on a tripod the PST is easy to align on the Sun – there is an internal "Sol Ranger" or sun spotting device. This consists of a small pinhole on the front of the PST body and a small opaque window on the top, near the eyepiece holder. When properly aligned on the Sun, the pinhole will let in light that will be projected onto the opaque glass in the form of a small bright ball. It is NOT necessary to put your eye up to the opaque glass. Best alignment will be found when this ball of light is near centre. Note: Adding a Solarmax 40 filter on to the front of the telescopes objective lens when double-stacking will obstruct the SolRanger.

The Coronado PST is equipped with a tuning ring that allows the user to adjust performance of the solar filter. The minute spectral adjustment of the etalon filter is accomplished by simply rotating this ring in either direction (but once set this adjustment is rarely required). The scope also has a small focusing knob under the rear box of the instrument. Once the PST is appropriately focused, the edge of the solar disc will appear sharp. However, if, for example, the prominences are not seen after the telescope has been focused, the filter adjustment may be of help. But remember, prominences are not always present around the limb of the Sun (Fig. 6.8).

The image of the Sun will appear a deep red across the entire disc. Be aware that it can take time to train one's eye for H-alpha viewing. Allow a minute for your eye to get used to the light. You need to look carefully into the image to see detail on the Sun's surface. The PST has a "sweet spot" in the centre of the disc where the image is fully tuned and detailed.

The PST usually comes with a 20 mm Kellner eyepiece. This is an economical three-element eyepiece and produces a fairly narrow field of view. It is not the best optically but because the PST uses basically monochromatic light (single wavelength) the effects of chromatic aberration are not a problem. Coronado makes "Cemax" eyepieces especially for its range of solar telescopes. Cemax eyepieces are contrast enhanced (multicoated) for the Sun and provide better images than those seen with the standard eyepiece. The best Cemax eyepiece for use with the PST is the 12 mm size.

Fig. 6.8 A Coronado PST being used to observe the Sun.

However, the author has found a Meade 15 mm Super Plossl (Series 4,000) eyepiece just as good. Observers can experiment with different eyepieces to see what is best for their scope.

It is best to use an eyepiece that has a rubber cup attached. Higher magnification eyepieces do not necessarily produce better results (less than 50× is best). The author has found the use of a Barlow lens or Polarizing eyepiece filter to be of limited value. Many amateurs use a zoom eyepiece with their H-alpha telescopes with much success.

For anyone deciding to double-stack their PST with a SolarMax 40 filter, there is no need to send the PST back to the manufacturer for spectral matching of the filters (as in some scopes). Almost any SolarMax 40 filter can be matched to the existing filter of the PST by using the adjustment tuner of the PST once the telescope has been double-stacked. The advantage of double stacking is the narrower bandpass of <0.6 A. This will increase the amount of surface detail that can be seen but prominences tend to be harder to see with a double-stacked scope. The extra filter will also cost as much as the PST does.

It is best to start out with a single-stack PST before venturing into the double-stack models.

It is possible to photograph the Sun's image in a PST by simply holding a camera up to the telescopes eyepiece (afocal method or eyepiece projection) and experimenting with different exposures. The author has used a Compact Canon digital camera with success and a Canon EOS 1,000D digital SLR camera. The lens of the camera remains in place but has a UV filter for protection against scratching. The camera lens is held in place against the rubber cap of the eyepiece and centred. Because exposures are short it is not necessary to mount the camera to the telescope (the camera may be too heavy in any case). The camera should be set to manual control, and at 200–400 ISO. The author has found with a 15 mm eyepiece, exposures of 1/25–1/50 s for prominences and 1/500 s for surface detail are the best. It is not possible to record surface detail and prominences on the same single frame because different exposures are required for each. The advantage in using a digital camera is that pictures can be viewed at once and discarded if required. The tricky part is in obtaining focus with the camera – the author basically focuses his camera at a distant object and uses that setting – maybe with a little fine-tuning at the eyepiece. You need to try a range of exposures to find out what is best for your particular equipment. This does not take long because the results are virtually instant. It is best to take whole disc images rather than magnified ones.

Once photos of the Sun have been taken, they can be downloaded onto a computer for processing. The author uses iPhoto on a MAC computer to adjust images (for exposure, contrast, definition, colour tones, etc.). It is also possible to use programs such as Registax to stack a number of photos and combine them to make one good image (for example, a photo showing prominences with one showing surface detail).

Occasionally when taking digital photos in H-alpha light, unwanted reflections and colour aberrations occur on images of the Sun. Sometimes odd effects occur on the printed photos. Differences also occur between one camera shop and another or from one printer to another (if printing your own photos). Again, a bit of experimenting can overcome such problems (Figs. 6.9 and 6.10).

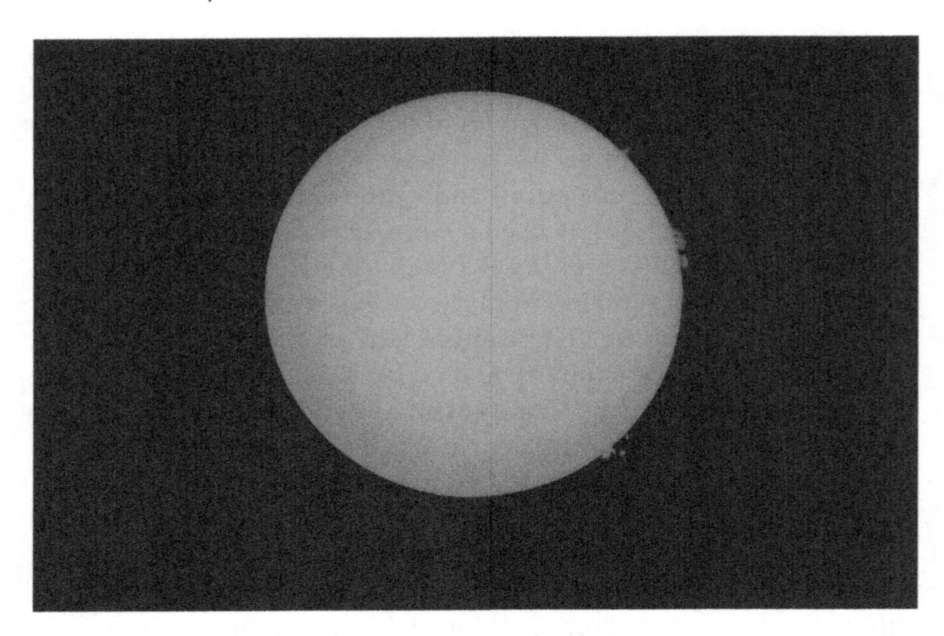

Fig. 6.9 Whole disc image of the Sun taken by the author using a Canon 1,000D digital SLR camera held to the eyepiece of a PST. Exposure was 1/25 s at ISO 200 on 22nd October 2010.

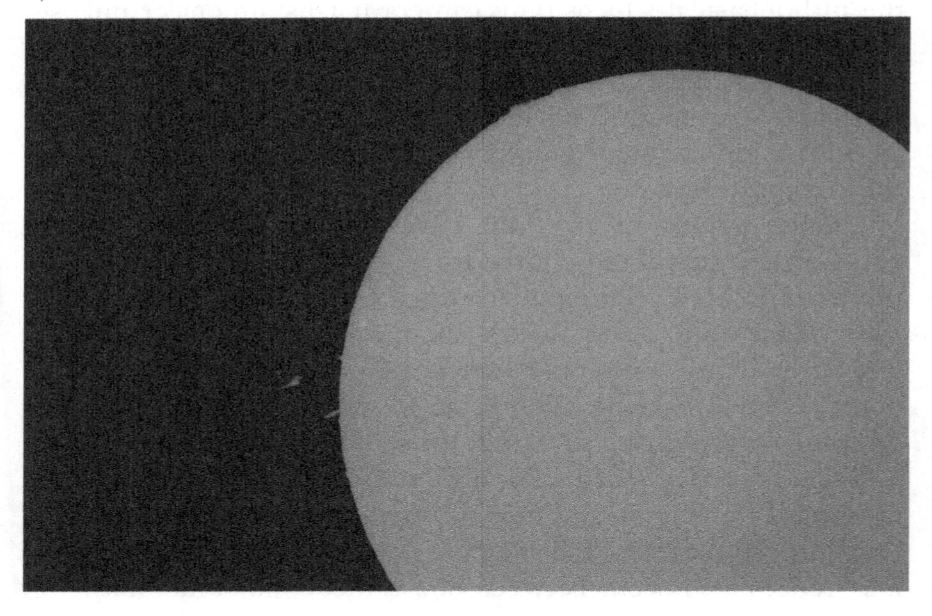

Fig. 6.10 Picture of prominences taken by the author via a Canon 1,000D digital camera held to the eyepiece of a Coronado PST on 28th January 2011. Exposure was 1/50 s at ISO 400. There is a filament on the disc near the prominence and a bright spot just above the filament.

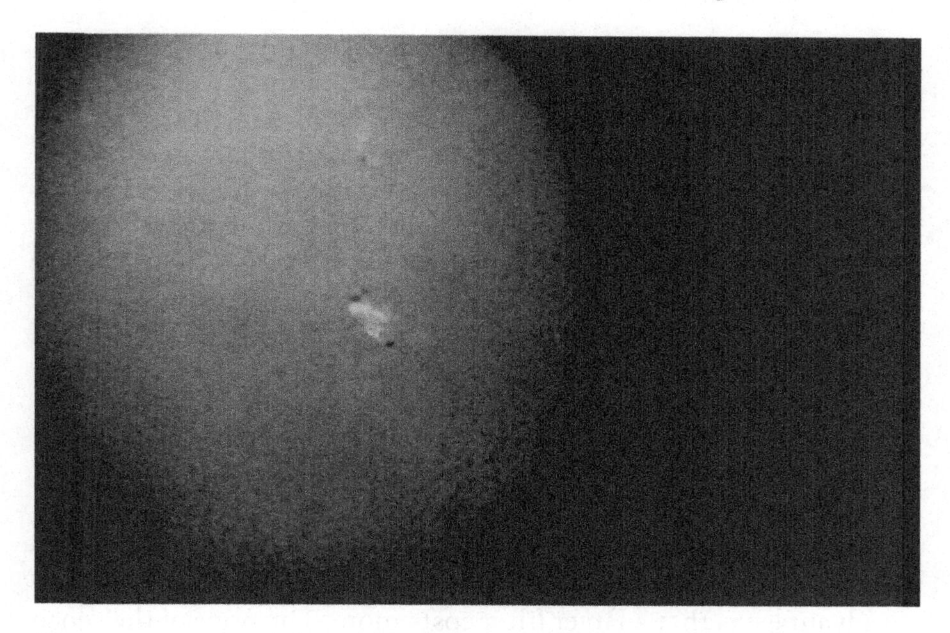

Fig. 6.11 Picture of the Sun's surface taken by the author on 10 March 2011 using Canon 1,000D digital camera held to the eyepiece of a Coronado PST. Exposed for surface detail – a sunspot group can be seen and plage that developed into a X1.5 solar flare (*bright* area between the three spots). Notice the mottled appearance of the surface. Exposure was 1/500 s at ISO 400.

Many solar observers take pictures of the Sun via a web cam, CCD camera or video camera mounted in place of the eyepiece. The results using such cameras are often better than digital cameras, because many images can be taken and stacked using image processing software (Fig. 6.11).

Coronado SolarMax II 60 mm

In 2010, the new Coronado SolarMax II 60 mm Solar Telescope (previously the MaxScope 60) became available to amateur astronomers who are more serious about solar observing. The scope has a larger aperture than the PST and a lower bandpass (<0.7 A) for observation of surface detail and prominences on the Sun in H-alpha light. The scope is particularly useful for serious amateurs who are into imaging. According to Coronado their new SolarMax II models represents a breakthrough in solar observing with the new and revolutionary RichView tuning assembly. This patented system allows direct tuning of the primary etalon filter. The manufacturers claim no other commercially available H-alpha telescope can provide the tuning range and accuracy of the SolarMax II. The scope can be tuned for the highest contrast views of active regions, flares, filaments, and other surface detail, or be quickly and easily re-tuned for prominences on the solar limb. Price is around US$1,500 (Table 6.2).

The SolarMax II 60 mm comes with one of three different sized blocking filters. There is a 5 mm or BF5 (similar to that in the PST), a 10 mm (BF10, recommended) and a 15 mm (BF15). What's the difference? A larger blocking filter gives you more blank space around the Sun and allows for further magnification before the image hits the edge of the field of view. Eyepieces with larger field stop diameters can be used without any vignetting in the field of view. A larger blocking filter is more suitable for imaging but the disadvantage is that a larger filter costs more. The price of the scope depends on which blocking filter is ordered. The author recommends the BF10 (Fig. 6.12).

The scope has a Sol Ranger finder (to line up the Sun), a clam-shell with three 1/4-20 holes for mounting, a blocking filter inside the diagonal, a 25 mm CEMAX eyepiece, a standard Coronado draw-tube-helical focuser arrangement and a travel-storage case. The etalon is mid-mounted at the end of the main tube and is adjusted via a tensioning screw or stick mounted on the side of the tube. As with the PST, the best eyepiece is the 12 mm Cemax eyepiece. It is worth trying other eyepieces. Note: The image in the 60 mm SolarMax is inverted in the N–S plane only compared to the image in the PST (E–W is the same).

Table 6.2 Specifications of the SolarMax II 60 mm scope

Aperture	60 mm
Focal length	400 mm
Focal ratio	f/6.6
Bandwidth	<0.7 A
Weight	3.0 kg (6 lb)
Standard eyepiece	25 mm Cemax
Eyepiece size/magnification	25 mm = 16×, 15 mm = 27×, 12 mm = 33×, 10 mm = 40×
Blocking filter	10 mm (BF10)

Fig. 6.12 A SolarMax II 60 mm scope being used to view the Sun in H-alpha light (Courtesy of Stephen Ramsen).

The extra weight of the 60 mm SolarMax (compared to a PST) means you will have to use a stronger mount. The 60 mm has more weight on the eyepiece side than the lens side.

This single etalon scope easily visually outperforms earlier Coronado double-stacked scopes. It provides excellent images and shows more surface detail and reveals smaller prominences. The larger aperture means higher magnification can be used (compared to the PST). The author has also found that a polarizing filter or neutral density filter in the eyepiece will reduce glare and show more surface detail (but they will reduce the brightness of prominences). A barlow lens can also be used with the 60 mm, while it had little value on the PST. With a double stack filter the bandpass comes down to about 0.5 A. Double stacking improves the surface detail but again the cost is much higher.

Other Coronado H-alpha Telescopes

There are a number of Coronado "MaxScope" solar telescopes that have been used by amateurs and professionals over the years.

Newer models include the RichView tuning system and are marketed under the name of "SolarMax II" scopes. The largest scope in the Coronado range is the SolarMax II 90 (previously the MaxScope 90). This is a professional scope with an aperture of 90 mm, focal length of 800 mm, focal ratio of f 8.8, and bandpass of 0.7 A. It is much larger and less portable than those previously discussed and it is much heavier, and requires a much more solid mount. It comes with a BF15 or BF30 blocking filter. Views however are superb.

Lunt 35 mm Solar Telescope

Lunt Solar Systems is a manufacturing and sales facility located in Tucson, Arizona. Lunt has a range of H-alpha telescope models, the first of which starts at just under US$500 but rise all the way through to the largest model at over US$8,000. Two of the most popular models for amateurs are discussed below.

The Lunt LS35THa is the most compact H-alpha telescope available. It has an aperture of only 35 mm and has a front mounted etalon with a bandpass of <0.8 A. Prominences and some surface detail can quickly be viewed through this very portable single stack system (Fig. 6.13).

The Lunt 35 scope comes with a 6 mm blocking filter (B600), a 1/4-20 camera tripod thread and a Televue Sol Searcher (finder). The scope has a small brass tuning wheel and a helical focuser (Table 6.3).

The advantages the scope has over the Coronado PST are:

- Larger externally mounted etalon with no central obstruction
- High quality 70° field of view eyepiece
- Rotatable blocking filter/diagonal allowing views from any angle
- Brighter image than PST
- Ability to balance the scope by moving it in its mounting rings
- Double stacking etalon available

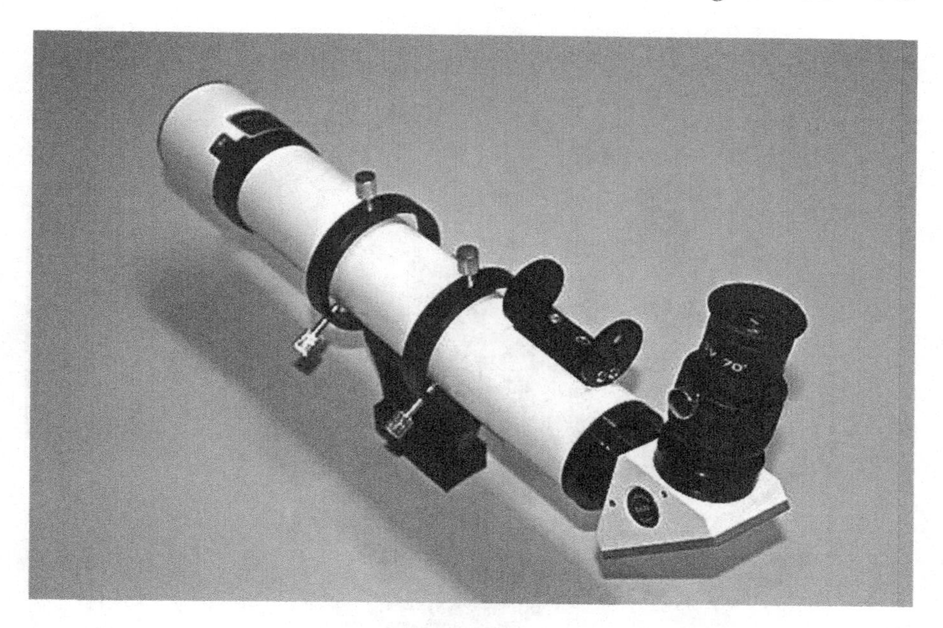

Fig. 6.13 The smallest Lunt H-alpha telescope is a good entry level one with aperture 35 mm (Credit: Lunt Solar Systems).

Table 6.3 Specifications of Lunt LS35THa

Aperture	35 mm
Focal length	400 mm
Focal ratio	f/11.4
Bandwidth	<0.8 A
Weight	1.8 kg (4 lb)
Standard eyepiece	10 mm
Eyepiece size/magnification	20 mm = 20×, 15 mm = 26×, 10 mm = 40×

Lunt LS 60THa

The Lunt LS60THa is a complete solar telescope with an objective lens of aperture 60 mm. The focal length is 500 mm with a 6 mm blocking filter inside the diagonal (a 12 mm blocking filter BF1200 is available as an option for imaging). Focus is achieved with a Crayford style focuser with a 10:1 reduction. An internal etalon with fine adjustment allows for a <0.8 A bandpass. The scope is the competitor of the Coronado SolarMax 60 (Fig. 6.14).

The scope offers slightly better viewing than the Coronado PST but a little less than a double stacked PST. Price is around US $1,300 (Table 6.4).

Fig. 6.14 Three solar telescopes in one! The *left* scope is a Lunt LS60 THa with pressure tuner (on the mid-side). The *right* scope is a 60 mm Lunt LSTHa with a standard tuning ring. The *middle* scope is a 90 mm Coronado SolarMax II (Courtesy of Stephen Ramsden).

Table 6.4 Specifications of Lunt LS60THa

Aperture	60 mm
Focal length	500 mm
Focal ratio	f/8.3
Bandwidth	<0.8 A
Weight	3.5 kg (7.7 lb)
Standard eyepiece	10 mm
Eyepiece size/magnification	20 mm = 25×, 10 mm = 50×

The LS60THa solar telescope is now available with the new air-pressure tuning system. The manufacturers claim this Pressure Tuner (PT) allows a better etalon adjustment than other systems. Apart from that the equipment is the same like the normal LS60THa telescopes. The Pressure Tuner telescopes are available in the different versions, with a B600 blocking filter for visual observing the sun, or a B1200 blocking filter for visual observing and imaging, and with a precious Feather Touch focuser.

Lunt Solar Systems also have a Pressure Tuned Hydrogen-alpha Telescope with a 80 mm singlet lens. An internal etalon system provides a bandpass of <0.7 A. The Lunt LS80THa telescope has a focal length of 560 mm, and includes a 12 mm blocking filter. Price is around US$3,000.

Other brands of solar telescopes and filter systems.

There are a number of other brands of solar telescope manufacturers, such as the Solarview range from Solarscopes in the UK. Solarview scopes have 50 and 60 mm aperture refracting solar telescopes with bandpass around 0.7 A.

Most manufacturers (e.g. Daystar, Solarscopes, Coronado, Lunt) also make separate filter systems that attach to your own refracting telescope (optical tube assembly or OTA). Such systems come in two main parts – a front mounted filter (that screws into the telescopes objective, may require an adaptor ring) and a blocking filter/diagonal for the eyepiece end. The advantage of these systems is that you can still use the telescope for night observation by removing the solar filters. However, a dedicated solar telescope would normally be the first choice for an amateur astronomer getting into solar astronomy. Dedicated telescopes are now cheaper than buying a separate filter system and adapting it to your own telescope. If buying a separate filter system to build your own scope – please get the advise of an experienced person to set it up – remember its your eyes!

Calcium II K Solar Observing

Observing the Sun in Hydrogen alpha light is the main method used by amateur solar astronomers, however it is not the only wavelength that can be used. The next most useful wavelength is that of the K line of ionised Calcium (Ca K line). This wavelength is centred on 3,933 A (393.3 nm) and is close to the boundary between visible light and the near ultraviolet. This reveals a lower, cooler region of the chromosphere than hydrogen alpha light. The Ca K line is also very sensitive to the presence of magnetic fields. If magnetic fields are present, absorption is less (more light is transmitted), with weaker magnetic fields showing as darker areas, and stronger magnetic fields as bright areas. Very strong magnetic fields, such as those in a sunspot appear very dark.

Views of the solar disc in Ca K light have a blue-purple colour. You can see a network of ionised calcium emissions that result from solar cells sweeping across through the magnetic fields. You can also see plage, active regions and magnetic storms. But Ca K light does not show up prominences nor filaments as clearly as hydrogen alpha light. Also some older people may not be able to see details in this wavelength of light (as its close to UV), hence Ca K telescopes are best suited for photographic/imaging use.

In Ca K light, views of the penumbra and umbra of sunspots appear similar to that in white light.

However you may see a bright ring or area of brightening around the penumbra of sunspots. One theory put forward to explain this phenomenon is that it is caused by the missing 10% of energy due to suppression of convective energy transport by underlying magnetic fields in the spot.

Intranetwork bright points, cell flashes or cell grains, originate exclusively within cell interiors in quiet areas of the solar surface. These are observed almost exclusively in the Calcium H (396.85 nm) and K lines. They are intermittent localised brightenings that last less than a minute and often re-appear a few times at 2–4 min intervals at about the same place, and frequently occur in pairs.

Observers should note that Ca K imaging and viewing are not as tolerant of high magnification as H-alpha or white light viewing. You also need longer exposures for imaging and lower magnification for visual use and afocal photography.

Advice: Try before you buy. Viewing in Ca K light is fundamentally different to viewing in H-alpha light.

A number of manufacturers make Ca K solar telescopes. Most use a filter with a wide bandpass (much wider than in H-alpha telescopes). Coronado makes a number of Ca K solar telescopes with different apertures from 40–90 mm. There is a 40 mm Ca K PST solar telescope with a filter centred on 3,934 A with a bandpass of 2.2 A (gives a range of 3932.9–3935.1 A). The Ca K PST scope is smaller than its H-alpha counterpart, and it uses a glass multicoated interference filter instead of an etalon. Ca K telescopes are usually much cheaper than H-alpha solar telescopes

Table 6.5 Specifications of the Coronado PST Ca K telescope

Aperture	40 mm
Focal length	400 mm
Focal ratio	f/10
Bandwidth	<2.2 A
Weight	1.5 kg (3 lb)
Standard eyepiece	20 mm
Eyepiece size/magnification	20 mm = 20×, 12 mm = 33×

because they have a much wider bandpass. Coronado's Cemax eyepieces also work well in a Ca K PST (Table 6.5).

Coronado also makes a 60 mm Ca K filter set that can be adapted to your own refracting telescope. It is centred on 3,950 A and has a bandpass of 2 A. It consists of an objective mounted front filter and a blocking filter/diagonal for the eyepiece end.

The Lunt 60 mm Ca K solar telescope has a focal length of 500 mm and a bandpass of <2.4 A with a 6 mm blocking filter. A slide tube on the scope allows for coarse focus adjustment, and a Crayford focuser takes care of fine adjustments, with a 10:1 reduction.

Did You Know?

A number of professional observatories on Earth and space probes such as SOHO and STEREO, use an instrument called a Coronagraph to study the Sun. Such an instrument is worth reviewing here, so that the reader can appreciate its use.

The Sun's corona can be seen during a total solar eclipse, when the Moon passing in front of the Sun blocks the disc. The coronagraph is an instrument that basically blocks out the Sun's disc to produce an artificial eclipse. French astronomer Bernard Lyot invented the coronagraph in 1930. Lyot's coronagraph consisted of a refracting telescope with a highly polished and scatter-free objective lens 5.1 in. diameter and focal length 134 in. (see Fig. 6.16). The lens (A) formed an image of the Sun on a black metal disc (B), which was just larger than the solar disc. The central portion of the solar image was reflected sideways out of the telescope (K and K'). A small lens (C) behind the black disc forms an image on a diaphragm (D) and small screen (E). The rim of the diaphragm cuts out light diffracted by the edges of the first lens; the screen eliminates the light of the solar images formed by reflection at the surfaces of this lens. Behind the diaphragm and screen an objective lens (F), protected from diffused light, produces an achromatic image of the corona at (B). A red filter was also placed at the eyepiece end of the instrument.

All the parts of the Lyot coronagraph were enclosed in a wooden tube just over 16 ft long. For the scope to work properly, it was essential that it should be used at high altitude; the Pic du Midi station in the Pyrenees Mountains (altitude 9,410 ft) was used. The instrument revealed prominences, the upper chromosphere and inner corona (Fig. 6.16).

Coronagraphs in outer space are much more effective than the same instruments would be if located on the ground. This is because the complete absence of atmospheric scattering eliminates the largest source of glare present in a terrestrial coronagraph.

While space-based coronagraphs such as LASCO avoid the sky brightness problem, they face design challenges in stray light management under the stringent size and weight requirements of space flight. Any sharp edge (such as the edge of an occulting disk or optical aperture) causes Fresnel diffraction of incoming light around the edge. The LASCO C-3 coronagraph uses an external and an internal occulter with a system of baffles to block stray light.

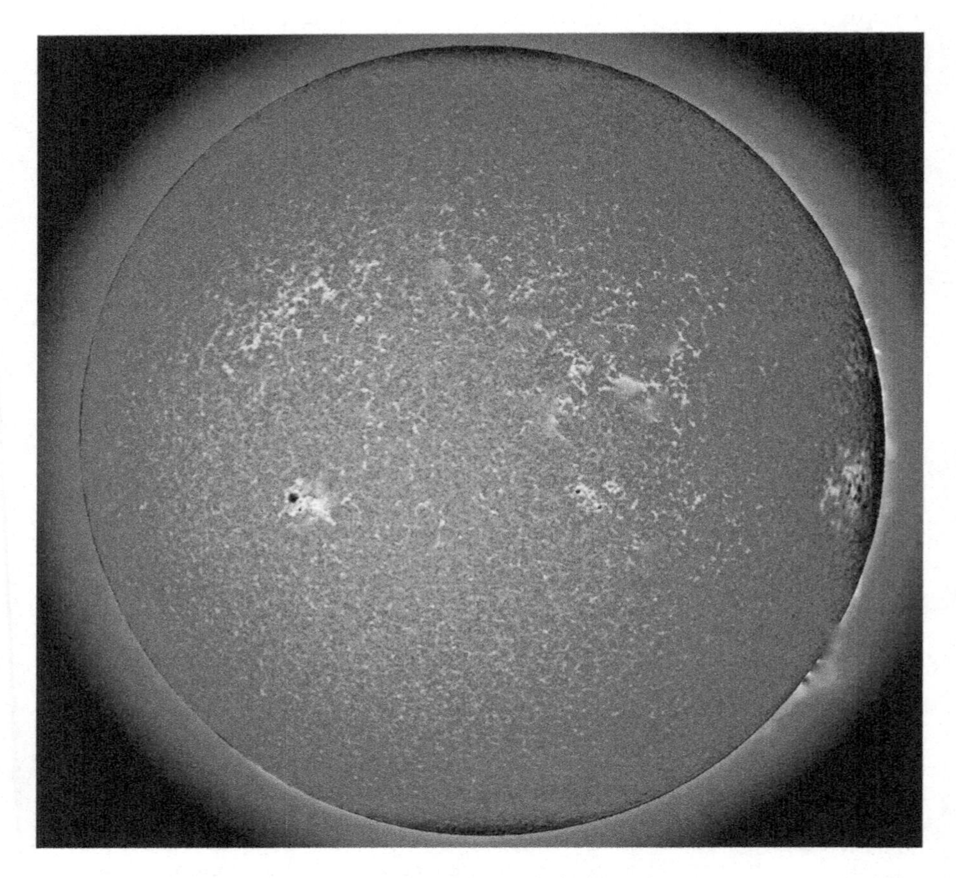

Fig. 6.15 Image of the Sun taken in Ca K light. The *white* areas are the plage. The *black* spots are sunspots (Credit: NASA/MSFC).

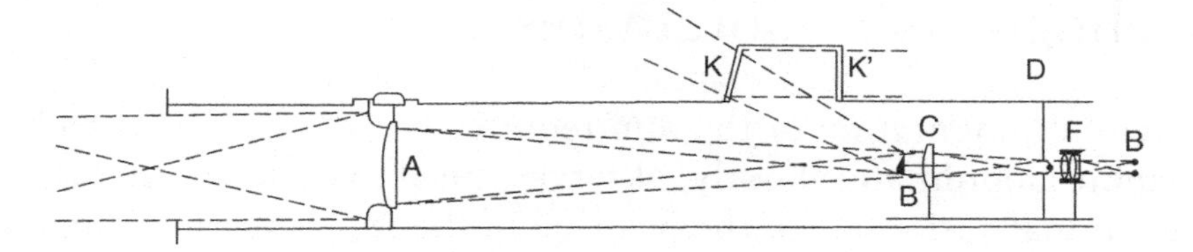

Fig. 6.16 Diagram showing the structure of the Lyot coronagraph.

Finding Directions on the Sun

Solar observers eventually need to be able to determine where the Sun's north and south pole are on the Sun's image, as well as east and west. A number of methods are available.

As the Sun appears in the sky, the northern half of the Sun will be toward the north celestial pole: the southern region faces the south celestial pole. Likewise the eastern limb of the Sun faces the eastern horizon of Earth, and the western limb faces the western horizon.

When viewing the image of the Sun through a telescope, things can get a little tricky as telescopes often invert images and/or switch left and right.

To roughly determine north, south, east and west we can use the following:

1. Through a stationary telescope, the western limb of the Sun will drift out of your view first, the eastern limb last. At right angles to the line between east and west is the north–south line.
2. The Sun rotates as seen from above its north pole in a counter-clockwise direction, east to west.

   As the Sun rotates, features disappear behind the western limb and appear at the eastern limb.
3. North and south on the Sun can be determined through a telescope by gently nudging the scope either north or south and noting the Sun's movement. For example, when you tap the scope towards the north, the southern hemisphere will start to leave your field of view.
4. If you make a record of the position of sunspots on paper, then compare your sketch with the orientation of the Sun on a website such as the SDO or SOHO sites, you can work out which way N, S, E and W are on your drawing. On satellite pictures N is up, S down, E to left and W to right.

Heliographic Coordinates

Accurate orientation of the Sun requires knowledge about Heliographic Coordinates. Firstly, observers need to be aware that there are annual variations in the position of the solar axis and equator. See Fig. 6.17.

The solar equator is inclined at an angle of 7.25° to the ecliptic (i.e. the plane of the orbit of Earth). Sunspots therefore do not appear to move in straight lines across the Sun's disc but in semi-elliptical paths. During parts of the year, the north pole of the Sun is tilted towards Earth, and at other times it is tilted away from Earth. The heliographic latitude Bo of the centre of the solar disc varies between +7.25° and −7.25°. When Bo is zero, in June and December, we are looking directly at the centre of the Sun's disc. Also variable with time is the position angle P between the solar axis and the north–south direction in the sky (P is + when the solar axis is inclined towards east, and – when inclined towards the west). P varies between +/−26.37°; when P is zero (January and July) the solar axis is aligned with celestial north–south. The two values Bo and P, from the date of an observation are all that is required to accurately fix the latitude of a feature on the Sun.

Solar astronomers measure solar longitude using the central meridian of the Sun. The longitude of the central meridian (Lo) is 0° at the beginning of each new solar rotation. Heliographic longitude is measured from east to west, and decreases by 13.2° per day.

The values of Bo, P and Lo are published yearly in Astronomy Yearbooks. See Fig 6.18.

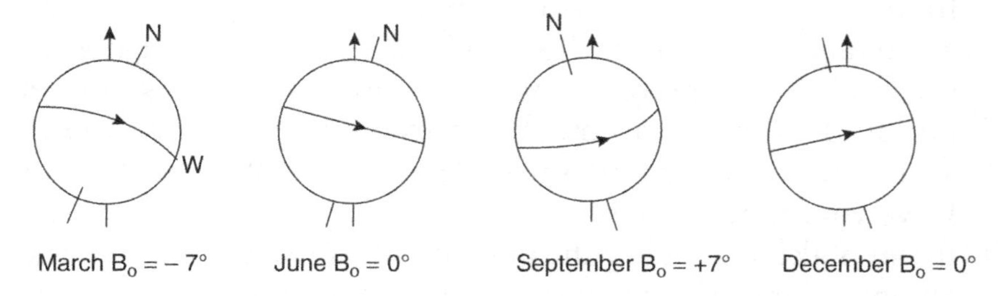

Fig. 6.17 Annual variations in the position of the solar axis and equator. Vertically up is to the celestial north pole.

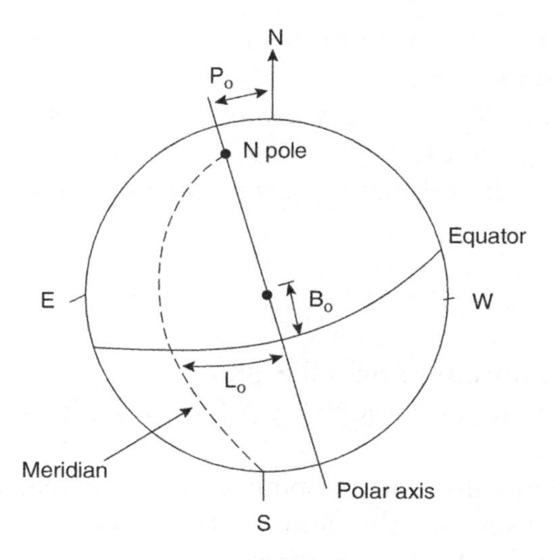

Fig. 6.18 System of heliographic coordinates.

Solar observing is carried out in conditions very different from that of evening observing. The bright light of the Sun needs to be treated with respect, especially when observation is through a telescope. However, commercially available telescopes are perfectly safe when operated within the manufacturers instructions and the amateur solar astronomer can gain much enjoyment from using a solar telescope, whether it is a white light type, a H-alpha type or a Ca K type. All three instruments provide a different experience and view of the Sun. Much fun can also be had when photographing or imaging the Sun.

Web Notes

For more information about the history of solar instruments see: http://www.europa.com/~telscope/solartele.doc
For useful videos on various aspects of solar viewing go to the Astronomy Now website: http://www.astronomynow.com/mag/1007/TheSun.html
For information about Coronado solar telescopes see: http://www.meade.com/products_pages/coronado/coronado/php

For information about Lunt solar telescopes see: http://www.luntsolarsystems.com
For information about Solarview brand of telescopes see http://www.solarscope.co.uk
For reviews of solar telescopes see: http://www.solarscopereviews.com

Books

For further information on solar photography try the book: "The Sun and How to Observe it" by Jamey Jenkins, published by Springer 2009.

For further information about Coronado solar telescopes see the book: "Observing the Sun with Coronado Telescopes" by Philip Pugh, published by Springer 2007.

7. Satellite Images of the Sun

A lot of information about the Sun has originated from space probes or satellites that orbit Earth or are in orbit between the Sun and Earth. These probes were discussed in Chap. 2. With advances in computer technology, images provided by these satellites are now posted daily on the internet and these images are available for use by professional and amateur solar astronomers, thanks to institutions supported by NASA and ESA. This chapter specifically focuses on satellite images and how to interpret them. The idea is that amateurs can use these images to support their own observations and identify various features of the Sun's active regions. Such images can also be used to view the Sun when it is cloudy here on Earth.

SOHO Satellite Images

SOHO (Solar and Heliospheric Observatory), is a project of international cooperation between ESA and NASA to study the solar wind, the Sun's core and outer corona.

The SOHO satellite was launched by NASA from the Cape Canaveral Air Station (Florida, United States) on 2nd December 1995. SOHO moves around the Sun in step with the Earth, by slowly orbiting around the first Lagrangian Point (L1), where the combined gravity of the Earth and Sun keep SOHO in an orbit locked to the Earth-Sun line. The L1 point is approximately 1.5 million kilometres away from Earth (about four times the distance of the Moon), in the direction of the Sun.

SOHO is operated from NASA's Goddard Space Flight Center (GSFC) near Washington. There an integrated team of scientists and engineers from NASA, partner industries, research laboratories and universities works under the overall responsibility of ESA. Ground

J. Wilkinson, *New Eyes on the Sun*, Astronomers' Universe,
DOI 10.1007/978-3-642-22839-1_7, © Springer-Verlag Berlin Heidelberg 2012

control is provided via NASA's Deep Space Network antennae, located at Goldstone (California), Canberra (Australia), and Madrid (Spain).

Images taken from SOHO are posted daily on the SOHO website: <sohowww.nascom.nasa.gov> and these can be easily and freely accessed by amateur astronomers. Amateurs can monitor these images if their own observations are restricted by lack of Sun or equipment. It is also useful for amateurs to compare their own observations with those of SOHO.

The Home page of the SOHO website has four main boxes on the right hand side – The Sun Now, Sunspots, Space Weather and Estimated KP. The first two contain images of interest in this chapter. The last two are concerned with Space Weather and will be discussed in Chap. 8.

The Sun Now

If you click on the "The Sun Now" box, eight images will come up on your computer screen (see Fig. 7.1). The first row of four images is from the EIT (Extreme ultraviolet Imaging Telescope). The EIT images the solar atmosphere at several wavelengths, and therefore, shows solar material at different temperatures. The images on the web page from left to right are labelled: EIT 171, EIT 195, EIT 284, EIT 304 (the numbers are the wavelengths in Angstrom units). Each image is falsely coloured to aid in identifying each wavelength (see Table 7.1).

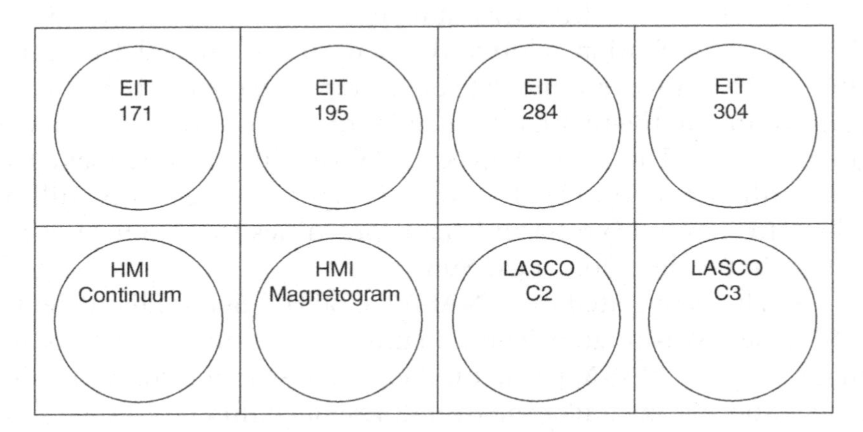

Fig. 7.1 Arrangement of images on "The Sun Now" section of the SOHO website.

Table 7.1 Details about EIT images on SOHO website

EIT number	Wavelength (A)	Temperature (°C)	Colour code
171	171	1 million	Blue
195	195	1.5 million	Orange
284	284	2 million	Yellow
304	304	60–80,000	Red

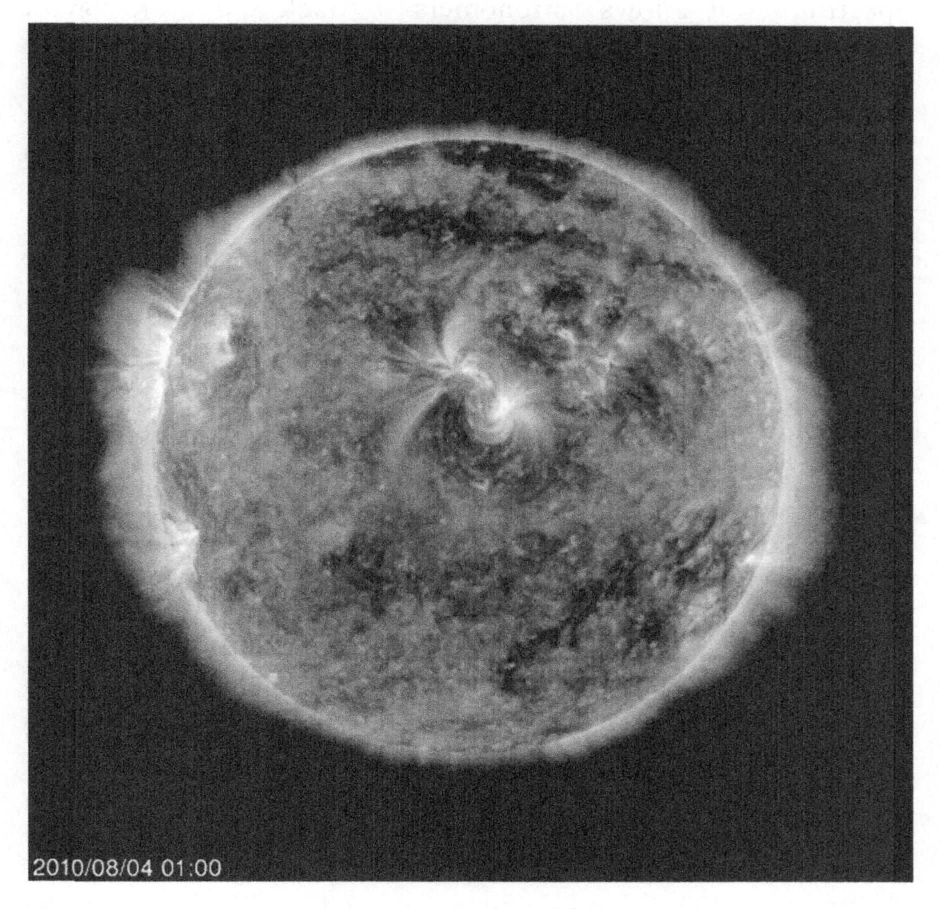

Fig. 7.2 EIT 171 image taken by SOHO on 4th August 2010. The *white* areas show the active regions. There was a large sunspot in the active region (*brightest white*) above *centre*.

For example, images marked "EIT 171" were taken at wavelength 171 A and the bright material corresponds to one million degrees. The hotter the temperature, the higher you look in the solar atmosphere (Fig. 7.2).

The first two images on the second row of "The Sun Now" are the MDI or HMI images. The MDI (Michelson Doppler Imager) images are taken in the continuum near the Ni I 6768 A line. A continuum image is formed by filtering parts of the visible light part of the spectrum. The most prominent features are the sunspots. The left hand image (coloured green) is the MDI Continuum and is very much how the Sun looks in the visible range of the spectrum – it allows astronomers to track the evolution of sunspots. The grey magnetogram image (on right side) shows the magnetic field in the solar photosphere, with black and white indicating opposite polarities. The black areas indicate magnetic fields pointing into the Sun (–), and the white areas denote fields pointing out of the Sun (+).

As of 1st January 2011, the MDI images have been replaced by HMI (Helioseismic and Magnetic Imager) images from the Solar Dynamics Observatory, but the older MDI images are available in the SOHO data/archives if you select a date prior to 1st January 2011.

Magnetograms are used to chart the magnetic fields running in and out of the Sun. They demonstrate that there is a lot of magnetism in the photosphere outside sunspots (Fig. 7.3).

The last two images in "The Sun Now" section of the SOHO website contains the LASCO images. LASCO (Large Angle Spectrometric Coronagraph) is able to take images of the solar corona by blocking the light coming directly from the Sun with an occulter disk, creating an artificial eclipse within the instrument itself. The position of the Sun's disc is indicated in the images by the white circle. There are two images presented – LASCO C2 (left) and LASCO C3 (right). C2 images (red) show the inner solar corona up to 8.4 million kilometers (5.25 million miles) away from the Sun. C3 images (blue) have a larger field of view: They encompass 32 diameters of the Sun. To put this in perspective, the diameter of the images is 45 million kilometers (about 30 million miles) at the distance of the Sun, or half of the diameter of the orbit of Mercury.

The images show the solar corona – bright spots on the images are bright stars behind the Sun.

The most prominent feature of the corona is usually the coronal streamers, those nearly radial bands that can be seen

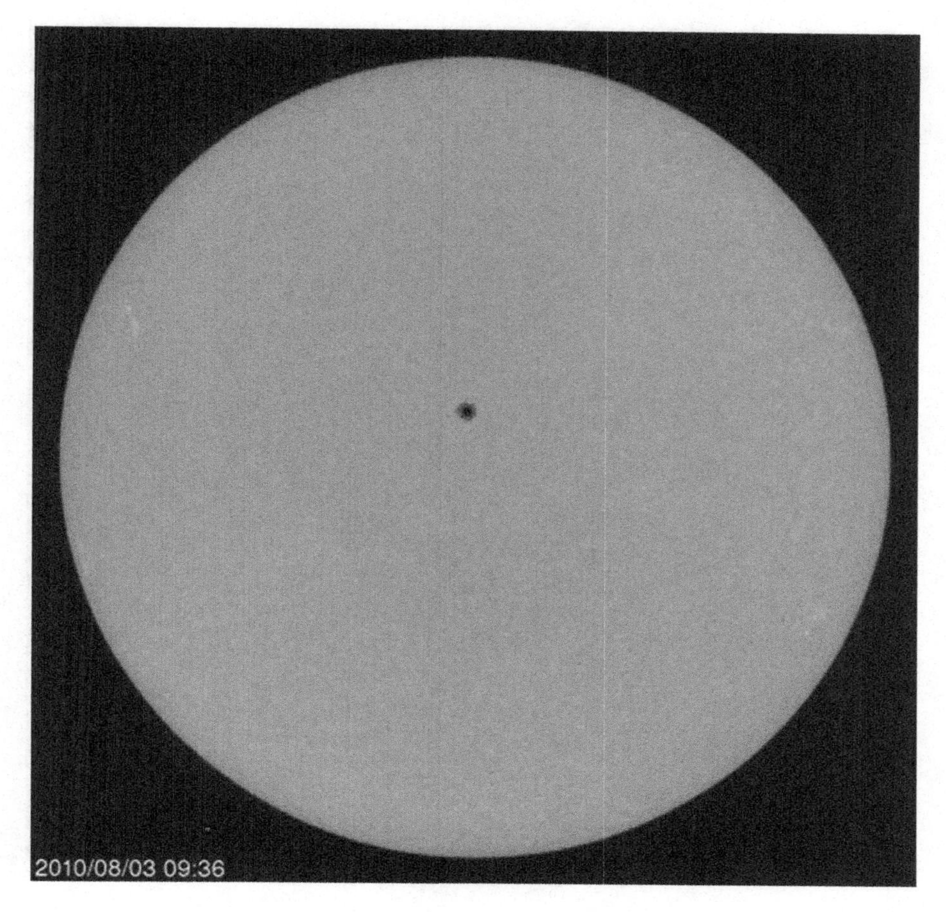

Fig. 7.3 MDI Continuum image of 3rd August 2010 from SOHO showing a large sunspot. The corresponding magnetogram for this image appears in Fig. 3.13.

both in C2 and C3. Occasionally, a coronal mass ejection can be seen being expelled away from the Sun and crossing the fields of view of both coronagraphs. The shadow crossing from the lower left corner to the center of the image is the support for the occulter disk (Fig. 7.4).

By clicking on each "The Sun Now" image, an enlargement is produced. Solar north is up in all images; east to left, west to right and south is down. Amateurs can use these images to identify

Fig. 7.4 LASCO C2 image taken on 4th August 2010 by SOHO. The *white* circle shows the position of the Sun's disc. The larger *red* disc is the occulting disc. The *white* radial bands are the coronal streamers.

N, S, E and W in their own photographs or in their own telescopes eyepiece.

Sunspots and Active Regions

Back on the home page of the SOHO website, there is a box on the right called "Sunspots". If you click on this, a black and white image of the current Sun is pictured, showing the sunspots visible and active regions. The date the image was taken also appears on the image. If you click on "List of available daily images" a list comes up for archived images for particular days.

An **active region** (AR) of the Sun is an area or event containing a sunspot, plage, facula or flare as a result of strong magnetic fields. Active regions in the corona are areas of enhanced density and temperature sometimes called coronal condensations. Other examples of active regions are areas on the photosphere where sunspots have faded, and x-ray bright points in the corona. Much of what astronomers observe in the photosphere is associated with active regions.

Since January 5th, 1972, astronomers have referred to each active region by a sequential four-digit number (e.g. 1074, 1075, ... etc.). The National Oceanic and Atmospheric Administration (NOAA) is delegated the responsibility of assigning an "AR" number to each new event. A typical name for a region might be AR1167. Sometimes solar events can last for several rotations of the Sun; in that case a region is given a different number for each appearance. AR numbers are usually limited to four digits up to 9,999. When the number 9,999 is reached (as it did on 14th June 2002, when AR 10,000 was observed) the four-digit sequence is often retained and AR10104 is referred to as AR0104.

Satellite images of the Sun found on the internet sometimes have AR numbers printed on them. Amateurs can compare their own observations with the internet images to determine what AR number a particular visible event has been assigned. The website of the Mees Solar Observatory in Hawaii <www.solar.ifa.hawaii.edu/mees.html>, provides daily active region maps as well as whole disc white light and monochromatic images; they also have an archive of earlier images.

Did You Know?

Scientists using the Michelson Doppler Imager (MDI) instrument on board the Solar and Heliospheric Observatory (SOHO) spacecraft have studied the sub-surface structure of selected areas on the Sun by analysing sound-generated ripples on its surface, using a technique similar to ultrasound diagnostics at a medical laboratory. They were able to construct a picture of magnetic structures inside the Sun because sound travels faster in solar regions with a strong magnetic field.

The scientists originally thought active regions had a simple structure. But instead of one large tube-like magnetic structure that rises from deep inside the Sun, the scientists found that active regions are made up of many small magnetic structures emerging at adjacent locations. Furthermore, the magnetic structures are replenished by others as they emerge, which makes the active region grow.

It is not yet known why a given region on the solar surface can suddenly erupt with magnetic structures and become active, or what causes the active region to be replenished by magnetic "reinforcements". According to the researchers, their data extends about 100,000 km (62,000 miles) inside the Sun – to the limit of the MDI – but the generation and storage of the magnetic structures probably occurs at the bottom of the Sun's convection zone at the tachocline, which extends another 100,000 km beneath the surface.

Another team of scientists from Stanford University explored the area beneath a sunspot and found it exhibited unusually pronounced rotation, spinning more than 200° counter-clockwise in less than 3 days. The team found a strong plasma vortex beneath the rotating sunspot and reported that the magnetic fields lacing the sunspot appeared twisted beneath the surface. Discovering the cause of twisted solar magnetic fields is important because it might eventually help predict stormy solar activity.

SDO Satellite Images

Images from the Solar Dynamics Observatory (SDO) can be seen on the website http://sdo.gsfc.nasa.gov/data/. Some of these images appear on other sites but this site allows you to click on each image and the wavelength is immediately given. For example, the first set of ten images are from the Atmospheric Imager Assembly (AIA) instrument and cover ten wavelengths from 94 A to 4,500 A. The second set of four images are composite images, whereby two or more wavelengths have been stacked on top of each other. The third set of images include a Magnetogram, Intensitygram and Dopplergram from the Helioseismic and Magnetic Imager (HMI) on SDO. Finally there is a Soft x-ray image from the Extreme Ultraviolet and Variability Experiment (EVE) and a couple of graphs. The graphs are of limited value to amateur astronomers but the images are useful, especially those near the visible region (4,500 A). Each image can be enlarged on the viewer's computer by double clicking (Fig. 7.5).

The SDO site also contains a gallery, news, pick of the week, hotshots, and movies. It is a very good site to explore.

Solar Monitor

The "Solar Monitor" is a website hosted at the Solar Physics Group, Trinity College Dublin and NASA Goddard Space Flight Centre's Solar Data Analysis Centre (SDAC). The pages contain

Fig. 7.5 When a rather large M3.6 class flare occurred near the edge of the Sun on 24th February 2011, it blew out a gorgeous, waving mass of erupting plasma that swirled and twisted for 90 min. NASA's Solar Dynamics Observatory captured the event in extreme ultraviolet light (304 A). Because SDO images are high definition, the team was able to zoom in on the flare and still see exquisite details (Credit: NASA/SDO).

near-real time and archived information on active regions and solar activity. The web site is located at www.SolarMonitor.org.html.

The website provides data from a variety of sources:

- SOHO: Data supplied courtesy of the SOHO/MDI and SOHO/EIT consortia. SOHO is a project of international cooperation between ESA and NASA.
- GONG: This work utilizes data obtained by the Global Oscillation Network Group (GONG) project, managed by the National Solar Observatory, which is operated by AURA, Inc. under a cooperative agreement with the National Science Foundation.
- NOAA: Solar Region Summaries, Solar Event Lists, GOES 5-min x-rays, proton and electron data from the Space Environment Centre, National Oceanic and Atmospheric Administration (NOAA), US Dept. of Commerce.

- SXI Full-disk x-ray images are supplied courtesy of the Solar X-ray Imager (SXI) team.
- XRT Full-disk x-ray images are supplied courtesy of the Hinode X-Ray Telescope (XRT) team.
- GHN Full-disk H-alpha images are supplied courtesy of the Global High Resolution H-alpha Network (GHN) team.
- STEREO Full-disk EUVI images are supplied courtesy of the STEREO Sun Earth Connection Coronal and Heliospheric Investigation (SECCHI) team.
- SOLIS Full-disk chromospheric magnetograms are supplied courtesy of the Synoptic Optical Long-term Investigations of the Sun (SOLIS) team.

Near the top-centre of the front page of the website you will find the date of the images (yyyymmdd) but not all are taken at exactly the same time. To the left of this are days before this date, while to the right are days forward of the date. A group of six images of the Sun at different wavelengths is also shown. If you click on "More instruments" under the six images you will find six more images (making 12 in total). See Fig. 7.6.

The first image is a black and white magnetogram of the Sun (from HMI, MDI or GONG). The second (yellow) is a visible wavelength image at 4,500 A from the AIA (Atmospheric Imaging Assembly) on SDO. The third image is in red hydrogen alpha light from GHN. The fourth image (blue) is taken at 174 A using the Fe IX/X line from SWAP or 171 A from AIA. The fifth image (brown) is from AIA at 193 A. The sixth image is from the X-Ray telescope on the Hinode spacecraft (XRT).

On the second page are six more images (we will refer to them as image 7–12). Image 7 is a black and white magnetogram (HMI or GONG). The eighth image from SOLIS is a chromospheric magnetogram and may be slightly different to images 1 and 7. Image 9 (pale yellow) from GONG shows a LOS magnetogram of the far side of the Sun. Image 10 (green) is from STEREO B and shows the Sun at 195 A using the Fe XII line. Image 11 (brown) is from AIA at 193 A using the Fe XII line. Image 12 (green) comes from STEREO A and shows the Sun at 195 A using the Fe XII line. See Table 7.2. Note: Sometimes the images in each box change

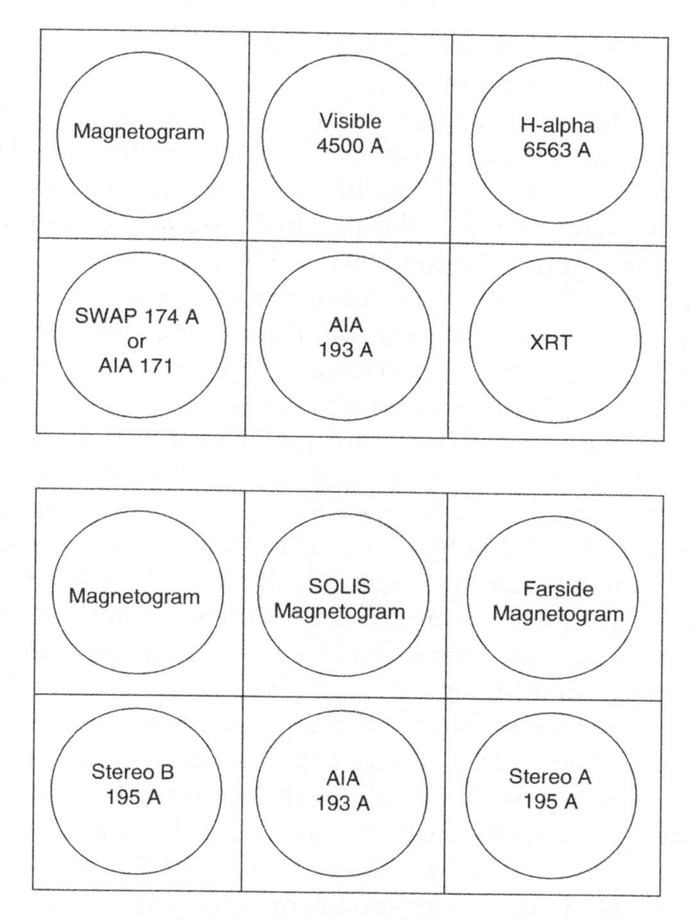

Fig. 7.6 Arrangement of images on the Solar Monitor website. The second set is found under the first set when you click on "More instruments".

Table 7.2 Guide to images on the Solar Monitor website

Image number	Wavelength (A)	Colour code	Source	Details
1	6,173	Black/White	SDO/HMI	Magnetogram
2	4,500	Yellow	SDO/AIA	Visible light
3	6,563	Red	GHN	H-alpha light
4	174	Blue	SWAP	Fe IX/X line
5	193	Brown	SDO/AIA	–
6	–	Deep red	Hinode/XRT	X-ray
7	6,173	Black/White	SDO/HMI	Magnetogram
8	–	Black/White	SOLIS	Chromospheric magn.
9	–	Pale yellow	GONG	Far side magnetogram
10	195	Green	STEREO B	Fe XII line
11	193	Brown	SDO/AIA	Fe XII line
12	195	Green	STEREO A	Fe XII line

their source from day to day and sometimes the images are missing because of technical problems.

Below the boxes/images on the web page is a "Summary" section that rates the activity level on the Sun for that date. For example, the activity level might be classed as "MEDIUM – 1M and 6C class flares in past 2 days". The Summary section will also list the "Most Active Region" by its NOAA number.

On the right hand side column there is a "Search" label. Clicking on it allows you to search the archives for past images – just by selecting a date or active region number.

Another useful feature of the Solar Monitor is the Active Region (AR) numbers. These are given as a five-digit number. This allows viewers to keep track of a particular AR over a number of days. There is also a heliographic grid overlaying each image (N up, S down, E left, W right). This grid can be used by amateurs to determine the latitude and longitude of particular AR on the Sun. Amateurs can also determine the AR numbers of features on their own photographs or observational drawings. Images are usually updated every 30 min (Figs. 7.7 and 7.8).

Below the image boxes of the Solar Monitor web page are details about the particular active region by number, titled "Today's NOAA Active regions". Listed AR details include (in order from left to right): Location (latitude and longitude), Hale (Mt Wilson sunspot magnetic polarity classification) and McIntosh sunspot classifications, Area, Nspots and Events (flares and class).

Did You Know?

The Naval Research Laboratory (NRL) group obtained the first x-ray image of the Sun in 1960 during a brief 5-min rocket flight. This primitive picture set the stage for the detailed x-ray images of the Sun taken from NASA sounding rockets in the early 1970s. During this time, solar x-ray instruments were also developed using NASA's series of small satellites known as the Orbiting Solar Observatories (OSOs), launched from 1962 to 1975.

Our understanding of the Sun changed forever when NASA launched the Skylab Earth-orbiting observatory on 14th May 1973. The observatory contained an x-ray telescope that was used to study the Sun's corona. Numerous x-ray images showed the Sun's x-ray corona was highly structured, containing coronal holes, loops and x-ray bright points. Magnetised loops mould, shape, and constrain the high-temperature, ionised gas in the corona. Electrons contained in coronal loops have temperatures of a few million degrees and a density of up to 100 million billion electrons per cubic metre.

Yohkoh, the Japanese-led international solar mission was launched 30th August 1991 from Kagoshima Space Centre in Japan. Yohkoh was the first spacecraft to

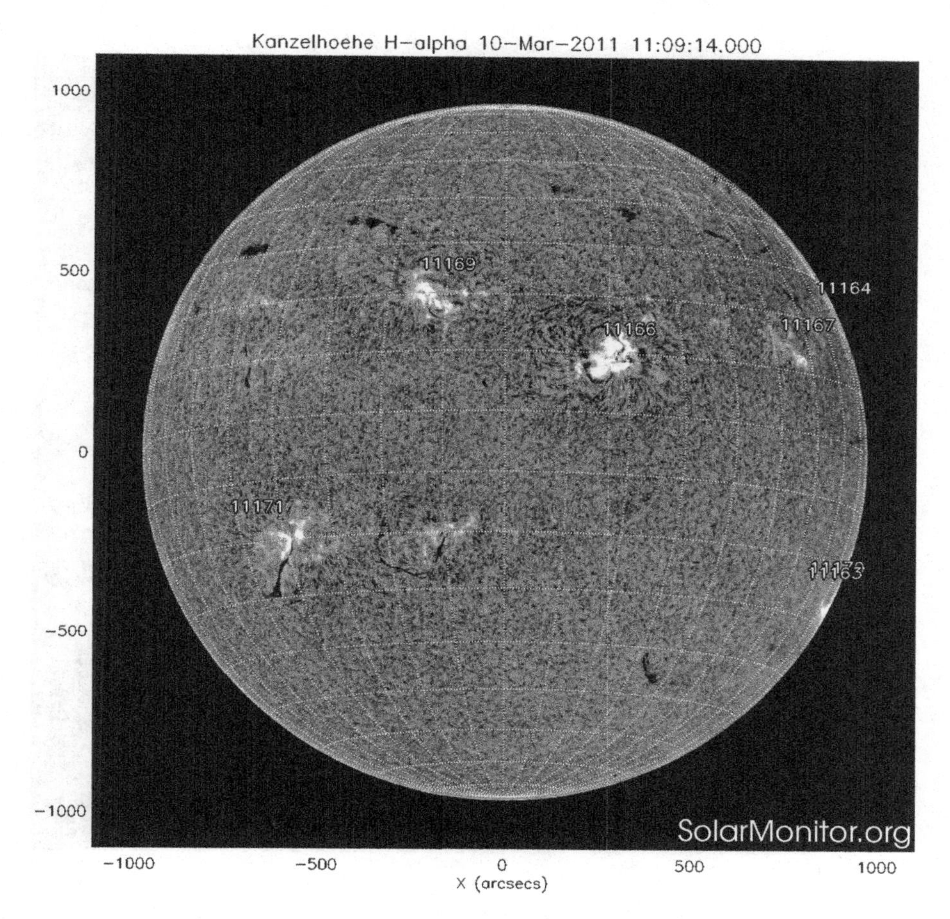

Fig. 7.7 H-alpha image of the Sun on 10th March 2011 showing the X1.5 flare in AR11166. Other ARs are also shown. Compare this with the author'sH-alpha image in Figure 6.11 (tilt the authors picture to get same orientation as shown here) (Image courtesy: SolarMonitor.org).

continuously observe the Sun in x-rays over an entire sunspot cycle, the roughly 11-year cycle in which the Sun goes from a period of numerous intense storms and sunspots to a period of relative calm and then back again. The value of the Yohkoh observations increased as the mission continued because they better revealed the many variable faces of the Sun. Additionally, the Yohkoh soft x-ray telescope (SXT) carried the longest-operating Charge Coupled Device (CCD) camera in space. After 10 years, the CCD camera – similar in operation to digital cameras now popular worldwide – is still taking beautiful x-ray pictures after collecting more than six million images.

Yohkoh observations have helped astronomers understand, better than ever before, how the Sun's magnetic fields are deformed and twisted, broken and

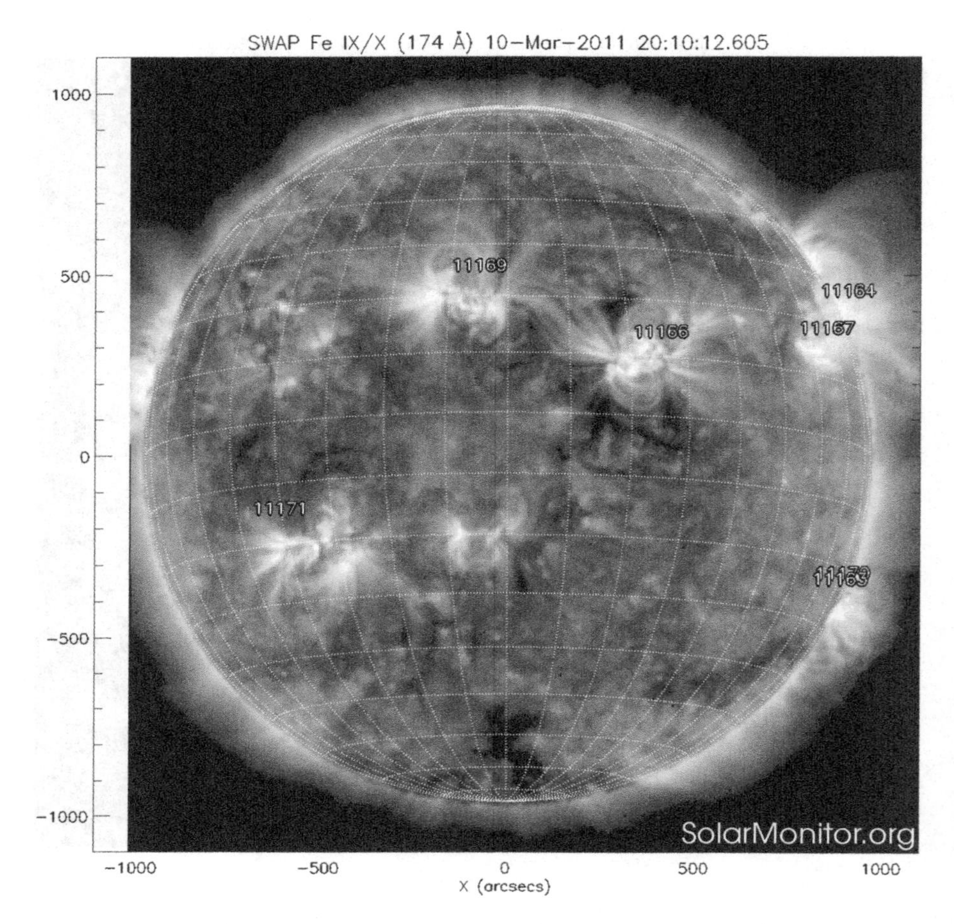

Fig. 7.8 SWAP image of the Sun on 10th March 2011 showing the X1.5 flare in AR 11166 – the wavelength however is 174 A (extreme-ultraviolet) so the image is different in appearance to the H-alpha image. The time of this image is slightly later than the H-alpha one (Image courtesy: SolarMonitor.org).

reconnected during flares; and how the electrified gas (plasma) of the Sun's corona is heated to millions of degrees during flares.

The Japanese Hinode spacecraft has an X-ray telescope (XRT) that provides an unprecedented combination of spatial resolution, field of view, and image cadence. It has the broadest temperature coverage of any coronal imager to date, from 1,000,000° to 30,000,000°. Its extremely large dynamic range permits detection of the entire corona, from coronal holes to the largest flares. Its high data rate permits observations of rapid changes in coronal magnetic and temperature structures. Results from Hinode have shown that the x-ray bright points, which appear all over the Sun, are actually small-scale magnetic loops (Fig. 7.9).

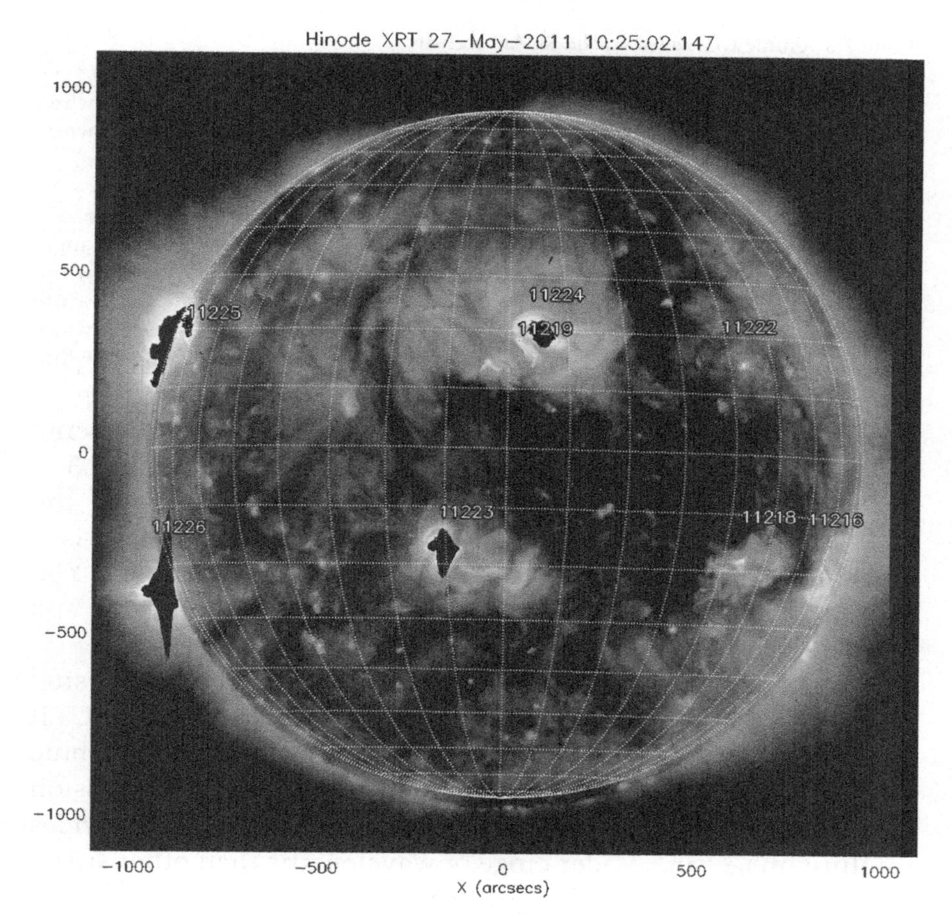

Fig. 7.9 X-ray image of the Sun taken on 27th May 2011 by the Japanese Hinode X-Ray Telescope. The most active region was AR 11226 and there was 3 C class flares at this time. X-ray images tend to be darker than those at other wavelengths (Credit: SolarMonitor.org/JAXA).

Solar Data Analysis Centre

Another useful website for satellite images in that of the Solar Data Analysis centre (SDAC) based at the Goddard Space Flight Centre. The website is <umbra.nascom.nasa.gov>. The first set of ten images comes from the AIA on SDO. The colour codes used to display AIA images in this website are the ones used to display similar bandpasses for SOHO EIT and STEREO EUV images.

Table 7.3 Guide to images on the SDAC website

Image	Wavelength (A)	Emission line	Colour code
1	94	Fe XVIII	Black/white
2	131	Fe XX	Black/white
3	171	Fe IX/X	Blue
4	193	Fe XII	Green
5	211	Fe XIV	Yellow
6	304	He II	Red/orange
7	335	Fe XVI	Black/white
8	1,600	Continuum	Black/white
9	1,700	Continuum/CIV	Black/white
10	4,500	Continuum	Black/white

Colour codes for the 94, 131, and 335 Å bandpasses were being developed at the time of publication of this book. See Table 7.3.

The SDAC website also contains two images from the Helioseismic and Magnetic Imager (HMI) on SDO. There is a magnetogram (black/white) and an Intensitygram (orange). You can also view a movie of the last 2 days of Magnetograms and Intensitygrams.

There are six images from other sources on the SDAC site: Hinode XRT, NSO SOLIS (He 10,830 A), NSO magnetogram (Ca II 8,542 A), RISE/PSPT (Ca II K), and HOA Coronagraph. The blue Ca II K image would be useful for amateurs with a Ca K emission line solar telescope. The advantage of this site is that it provides satellite images in a wider range of wavelengths than other sites.

Helioviewer and JHelioviewer

Helioviewer.org is a web-based site that provides solar images from SOHO and SDO from March 1993. The site is funded by ESA and NASA. You simply enter the date required and time, the observatory (satellite) will be either SOHO or SDO, and the instrument – AIA or HMI. The "Measurements" box contains a list of wavelengths (in Angstroms) for the various images available, these go from 94–335 A and 1,600–4,500 A. If you want an image of the Sun in a visual wavelength (white light) then select 4,500 A. Hints and instructions for use are available on the web page.

The program can put together a sequence of images for various lengths of time and provide you with a movie of the sequence. Alternatively you can download single images.

Note: Helioviewer makes use of Coordinated Universal Time (abbreviated UTC). UTC is the time standard by which the world regulates clocks and time. Computer servers, online services and other entities that rely on having a universally accepted time use UTC for that purpose.

JHelioviewer is new visualization software/website that enables everyone, anywhere to explore the Sun. Developed as part of the ESA/NASA Helioviewer Project; it provides a desktop program that enables users to call up images of the Sun from the past 15 years. More than a million images from SOHO can already be accessed, and new images from NASA's Solar Dynamics Observatory are being added every day.

Using this new software, users can create their own movies of the Sun, colour the images as they wish, and image-process the movies in real time. They can export their finished movies in various formats, and track features on the Sun by compensating for the solar rotation. However, the true power of the tool lies in its ability to allow cross-referencing of different aspects of the large data sets; many events observed on the Sun are interconnected and occur over vastly different temporal and spatial scales.

For the first time, JHelioviewer allows users to overlay series of images from the Sun, from different instruments, and compile an animated sequence, which they can then manipulate as they watch, in order to follow a solar event from start to finish. Existing tools provided the option to either see the large-scale corona, or small patches on the solar surface but there was no option to overlay the two and zoom in and out as you watch the Sun's activity evolve. JHelioviewer tackles this hierarchy of scales by coupling all solar scales: you have small-scale phenomena tying into large-scale events.

JHelioviewer is written in the Java programming language, hence the "J" at the beginning of its name. It is open-source software, meaning that all its components are freely available so people can help to improve the program. They can even reuse the code for other purposes; it is already being used for Mars data and in medical research. This is because JHelioviewer does not need to download entire data sets, which can often be huge; it can just choose enough data to stream smoothly over the Internet. It also allows data to be annotated, say solar flares of a particular magnitude to be marked, or diseased tissues in medical images to be highlighted.

The JHelioviewer software is available to download for several operating systems. The web-based image browser found at <Helioviewer.org> complements it. There is also a JHelioviewer handbook available for downloading. Video tutorials are also on the website as are full instructions for users.

STEREO 3D Images of the Sun

STEREO (Solar TErrestrial RElations Observatory) is a solar mission launched on 26 October 2006 by NASA. It consists of two nearly identical spacecraft, one orbiting ahead of Earth (A) and the other behind Earth (B). The two STEREO spacecraft reach 180° separation and observe the entire Sun simultaneously in the same wavelength. See Chap. 2.

Starting in February 2011, and continuing on for the next 8 years, STEREO has provided the first ever 360-degree view of the Sun. By combining images from the STEREO A and B spacecraft, together with images from NASA's Solar Dynamic Observatory (SDO) satellite, a complete map of the solar globe can be formed. Previous to the STEREO mission, astronomers could only see the side of the Sun facing Earth, and had little knowledge of what happened to solar features after they rotated out of view.

A 3D picture of the Sun cannot be reproduced on this page (only 2D), but you can view the image on the NASA/STEREO website at http://stereo.gsfc.nasa.gov/360blog/ (see example in Fig. 7.10). The coordinates of the STEREO images are determined and converted into heliographic maps. Only pixels that are inside of the solar limb are used, since unique heliographic coordinates cannot be calculated above the limb. This would be fine if all the emission was coming from the surface, but because the Sun has an atmosphere that extends well above the surface, vertical features near the limb tend to be projected nearer to the limb than their

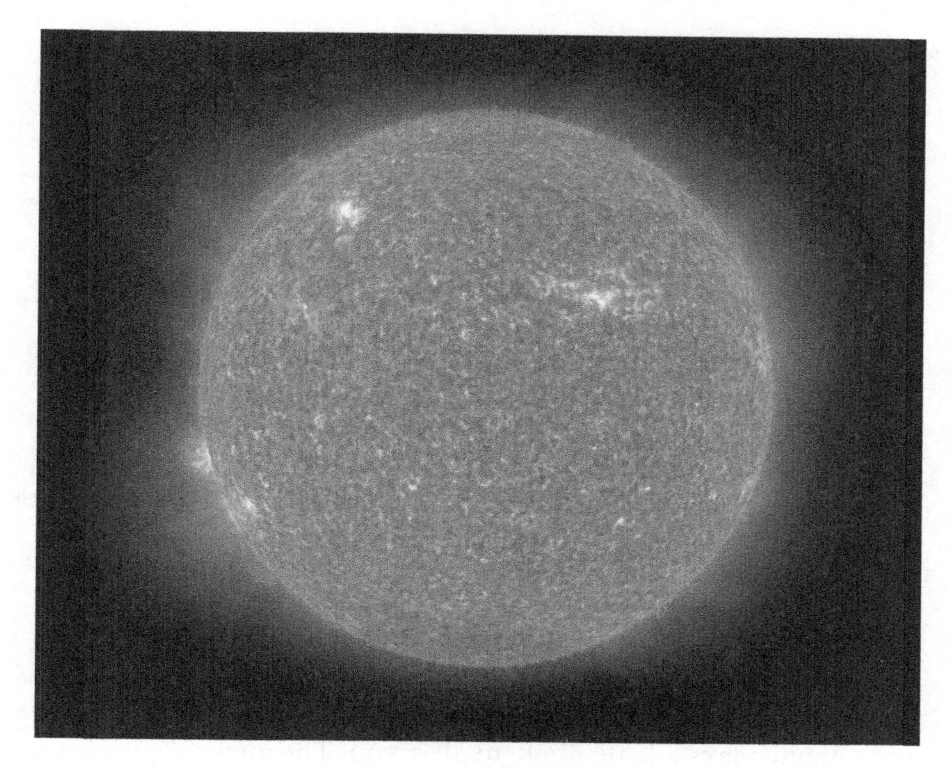

Fig. 7.10 A 3D picture taken on 4th January 2011 in the Helium II emission line at 304 Å (Credit: NASA/STEREO). See also Fig. 2.8.

actual heliographic position. In the heliographic maps this shows up as smearing near the edges of the map.

SDO observes the Sun in three of the four wavelengths seen by STEREO: in the Helium II emission line at 304 A, representative of plasma at about 80,000°; the Iron IX line at 171 A (1.3 million degrees); and the Iron XII line at 195 A (1.6 million degrees). SDO does not have a bandpass equivalent to the Iron XV line at 284 A (two million degrees) seen by STEREO, but the slightly cooler Iron XIV line at 211 A, is a reasonable substitute. SDO images are also corrected for differences in the Sun's rotation rate and combined with those of STEREO.

3D Sun for iPhone or iPod

A new iPhone application developed by NASA-supported programmers delivers a live global view of the sun directly to your iPhone, iPod or Smart phone. Users can fly around the Sun, zoom in on active regions, and monitor solar activity. The name of the application is "3D Sun" and it may be downloaded free of charge at Apple's app store. Just enter "3D Sun" in the store's search box or visit <http://3dsun.org> for a direct link. The real time images come from the STEREO spacecraft and are in the extreme ultraviolet (EUV) portion of the Electromagnetic Spectrum. There are four wavelengths available – 171, 195, 284, and 304 A each with its own false-colour. EUV is where the action is. Solar flares and new sunspots shine brightly at these wavelengths. EUV images also reveal coronal holes, vast dark openings in the Sun's atmosphere that spew streams of solar wind into the Solar System. There are also news items, details on the current condition of the Sun, as well as an SDO gallery (Fig. 7.11).

With this application, you can spin the Sun, zoom in on sunspots, inspect coronal holes, and when a solar flare erupts, your phone plays a little jingle to alert you. The application comes alive on its own when the Sun grows active or when interesting events happen. For example, a recent alert notified users that a comet just discovered by STEREO-A was approaching the Sun. When solar heating destroyed the comet, the application played a movie of the comet's last hours.

Recently, STEREO-B was monitoring a far side sunspot (AR1041) when the sunspot's magnetic field erupted. For the first time in almost 2 years, an active region on the Sun produced a strong "M-class" solar flare. The unexpected interruption of the Sun's deep solar minimum was invisible from Earth, but anyone with the 3D Sun had a ringside seat for the blast.

Satellite images of the Sun provide the amateur astronomer with another exciting method of observing the Sun. As satellite and instrumental technology improves so does the range of activities for amateurs. You don't even have to have a computer at hand – an iPhone, iPod or smart phone will do. This is truly an exciting time to monitor the Sun – our nearest star.

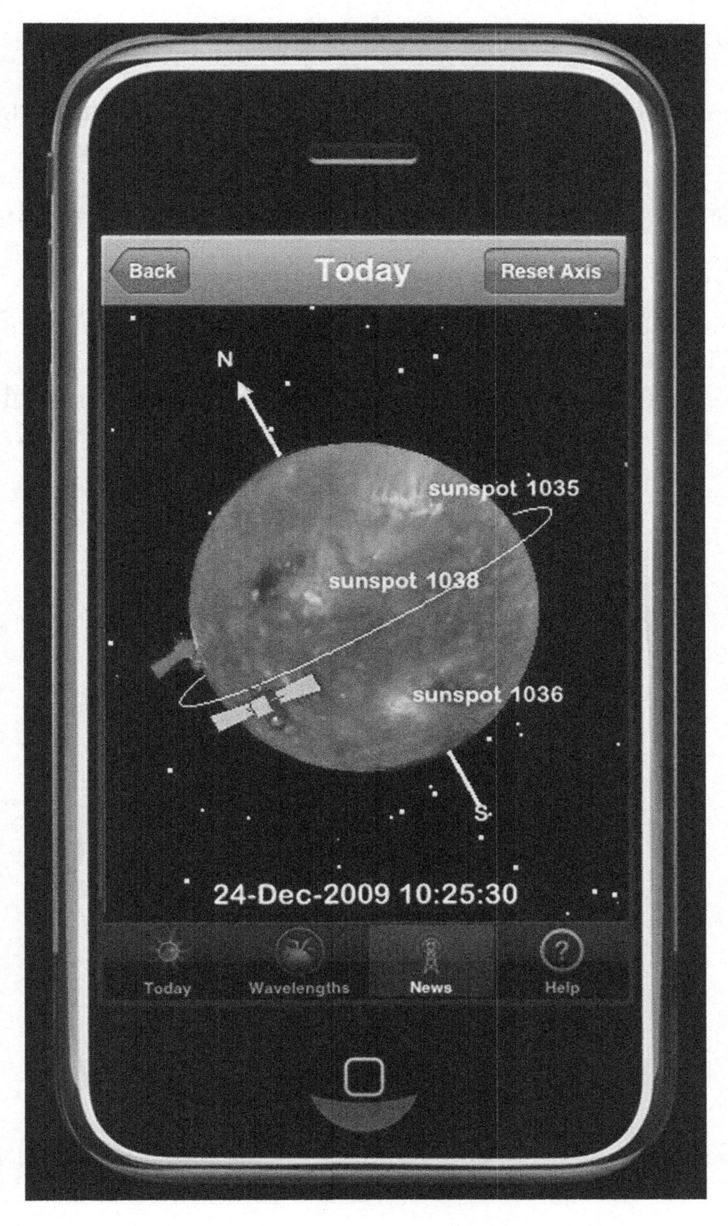

Fig. 7.11 Picture of the 3D Sun as seen on the cell or iPhone.

Web Notes

For more information about HMI visit: http://hmi.stanford.edu/
The SOHO website is http://sohowww.nascom.nasa.gov
The SDO website is http://sdo.gsfc.nasa.gov/data/.
The website is Solar Monitor is http://www.SolarMonitor.org.html.
The SDAC website is http://umbra.nascom.nasa.gov
For the Helioviewer website go to http://jhelioviewer.org or http://helioviewer.org.
For the Stereo 3D website: http://stereo.gsfc.nasa.gov/360blog/
Also try http://www.nasa.gov/topics/solarsystem/features/iphone-sun.html
The 3D Sun cell phone website is http://3dsun.org

8. Space Weather

Most of us are aware that the Sun influences the weather on Earth. The Sun also has a profound impact on our upper atmosphere and the magnetic field that surrounds our planet. Ejections of particles and energy from the Sun sometimes set off electrical disturbances in this magnetic field that affect sensitive communications and electrical distribution systems worldwide. The weather on Earth's surface is connected with conditions in the lower layers of the atmosphere. Space weather is something new, and has to do with the changing conditions in interplanetary space, the solar wind and the Earth's magnetosphere. This chapter will focus on Space Weather and we shall begin with the solar wind.

The Solar Wind

The extremely hot outer atmosphere of the Sun does not remain still. The Sun is continually emitting charged particles out into its surroundings. The realisation that something is always being expelled from the Sun came from observations of comets, auroras and geomagnetic storms. Comets appear unexpectedly almost anywhere in the sky, but they all seem to orbit the Sun in highly elliptical orbits and once close to the Sun, they grow tails that always point away from the Sun. A comet travels headfirst when approaching the Sun and tail first when departing from it. Ancient Chinese astronomers thought that there might be some force pushing the comets tail. In the early 1600s the German astronomer Johannes Kepler proposed that it was the pressure of sunlight that pushed the comets tail away from the Sun (Fig. 8.1).

In 1859, the British astronomer Richard Carrington suggested there might be a continuous stream of particles flowing outward

J. Wilkinson, *New Eyes on the Sun*, Astronomers' Universe,

DOI 10.1007/978-3-642-22839-1_8,

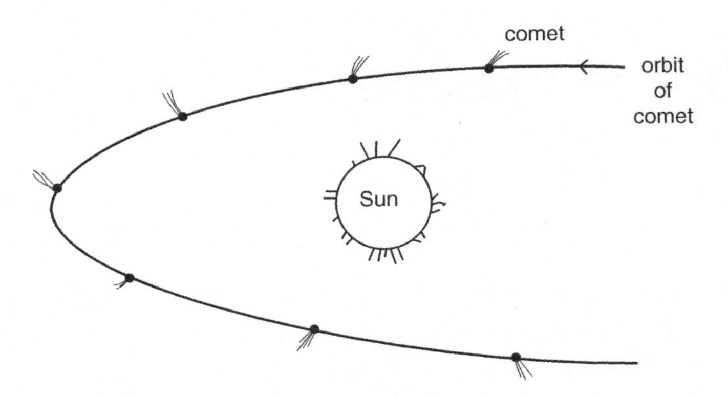

Fig. 8.1 Path of a typical comet as it orbits the Sun. Notice how the tail always points away from the Sun.

from the Sun. At this time, Carrington and Richard Hodgson independently made the first observation of what would later be called a solar flare. On the day following the flare, a geomagnetic storm was observed on Earth, and Carrington suspected that there might be some connection.

In 1916, the Norwegian physicist Kristian Birkeland suggested that auroral activity around Earth's polar regions was produced by particles from the Sun, and that these particles consisted of both negative electrons and positive ions.

In the 1930s, scientists observing an eclipse of the Sun determined that the temperature of the Sun's corona must be at least a million degrees because of the way it stretched out from the Sun.

In the early 1950s, the German astrophysicist Ludwig Biermann reported that streams of electrically charged particles poured out from the Sun continuously and in all directions at speeds of 500–1,000 km/s. He also thought that these solar particles caused the tail of a comet to always point away from the Sun.

In 1958, Eugene Parker of the University of Chicago, USA, developed a model in which there was a supersonic flow of charged particles from the expansion of the Sun's corona, called the solar wind. Parker also argued that this wind would pull the Sun's magnetic field into surrounding interplanetary space attaining a spiral shape because of radial flow and the Sun's rotation.

The American spacecraft, Explorer 1, launched in 1958, was the first to detect energetic charged particles in magnetic field surrounding Earth. These particles were found trapped in two donut-shaped magnetic fields called the Van Allen radiation belts (named after physicist James A. Van Allen, whose instruments discovered them). A ring current also circulates around the Earth just outside the Van Allen belts. Charged particles in the belts and ring originate from the solar wind and Earth's own ionosphere. See Fig. 8.5.

The first direct measurements of charged particles outside Earth's magnetic field were made by Soviet scientists using four ion traps on board the Luna 2 spacecraft launched to the Moon in September 1959. All four traps (distributed around the spacecraft) contained charged particles from the solar wind. Instruments on board the Explorer 10 spacecraft also carried out solar wind measurements in 1961.

All doubts about the existence of a solar wind were removed by measurements made by the Mariner 2 space probe, launched in 1962, while en route to Venus. The velocity of the solar wind was determined to average 500 km/s. The average wind ion density was shown to be five million protons per cubic metre close to Earth. The Mariner 2 data unexpectedly showed the solar wind to have a slow and fast component. The slow wind moves at 300–400 km/s and is twice as dense as the fast wind. The fast wind has a speed around 750 km/s. The high velocity wind swept past Mariner 2 every 27 days, suggesting it was associated with the rotating Sun. Furthermore, peaks in geomagnetic activity, also repeating every 27 days, were correlated with the arrival of these high speed streams at the Earth. This suggested a direct relationship exists between some source on the Sun and disturbances of the Earth's magnetic field.

The **solar wind** is known to consist of rarefied plasma or an electrically neutral mixture of electrons, protons, and heavier ions, and magnetic fields streaming radially outward from the Sun in all directions at supersonic speeds. These particles can escape the Sun's gravity because of their high kinetic energy and the high temperature of the corona.

In 1990, the Ulysses probe was launched to study the solar wind from a polar orbit around the Sun. All prior observations had been made at or near the solar system's ecliptic plane. Its instruments found that the solar wind blows faster at the poles than at the equatorial regions and that the fast wind is the dominant component.

In the late 1990s the Ultraviolet Coronal Spectrometer (UVCS) instrument on board the SOHO spacecraft observed the acceleration region of the fast solar wind emanating from the poles of the Sun, and found that the wind accelerates much faster than can be accounted for by thermodynamic expansion alone. Parker's model predicted that the wind should make the transition to supersonic flow at an altitude of about four solar radii from the photosphere; but the transition now appears to be much lower, perhaps only one solar radius above the photosphere, suggesting that some additional mechanism accelerates the solar wind away from the Sun.

Observations of the Sun between 1996 and 2001 showed that emission of the slow solar wind occurred between latitudes of 30°–35° around the Sun's equator during the solar minimum, then moved toward the poles as a period of solar maximum approached.

A comparison of Ulysses data with x-ray and white-light coronagraph data indicates that at solar minimum the fast wind escapes from the Sun along the open magnetic field lines of polar coronal holes. The slow wind originates from the equatorial region in coronal streamers, where the magnetic field lines are closed close to the Sun and oppositely directed and parallel further from the Sun (see Fig. 8.2). **Coronal streamers** are wisp-like stream of particles travelling outwards through the Sun's corona. They are visible during a total solar eclipse or in images taken with a coronagraph. Coronal streamers are thought to be associated with active regions and/or prominences and are most impressive near the maximum of the solar cycle. Although they can be longer than the diameter of the Sun, they are very tenuous; the material in them gradually moves away from the Sun and becomes part of the solar wind (see Fig. 8.3).

At a solar maximum the wind speeds are more variable and lower and they seem to come from a variety of sources including coronal holes, coronal streamers, coronal mass ejections and solar active regions. The large polar coronal holes shrink and disappear at maximum activity and the fast wind comes from smaller holes all over the Sun (Table 8.1).

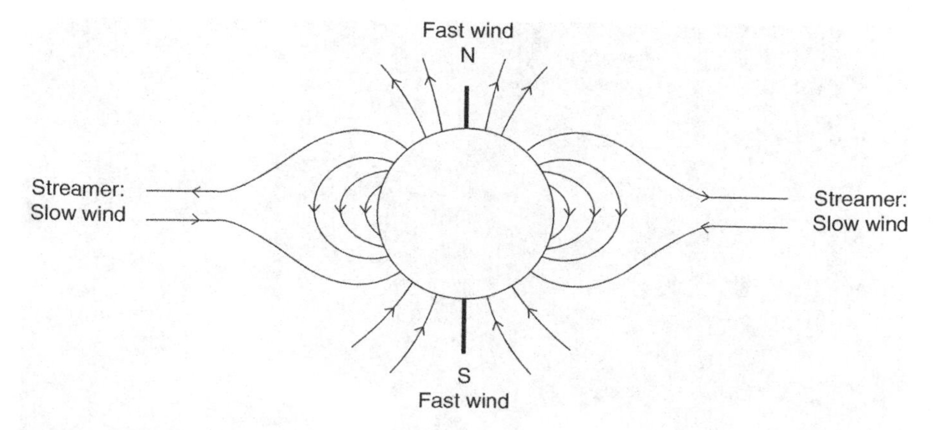

Fig. 8.2 Magnetic field lines around the Sun at a time of solar minimum. The fast wind escapes from the polar regions where the field is open, and the slow wind escapes from the equatorial regions where the field is more closed.

The fast and slow solar winds do not blow uniformly from all points of the Sun, but instead depend on latitude and upon the Sun's magnetic configuration (which varies with the 11 year sunspot cycle). The wind outflow begins slowly near the Sun, where gravity is the strongest, and then accelerates out into space.

Did You Know?

The solar wind can be described by considering it as electrically charged particles moving in a magnetic field. The wind's physical properties are variable in both space and time. Since the 1960s, these variations were thought to be due to waves produced by the Sun.

Hannes Alfven (1908–1995) received the Nobel Prize in Physics in 1970 for his work on magnetohydrodynamics. He showed that when charged particles are perturbed in the presence of a magnetic field, they will move up, down, and forward, oscillating and propagating like a wave. These oscillations are known as Alfven waves and their speed increases with increasing magnetic field strength and decreasing particle density. It has been proposed that Alfven waves might be able to explain how the expanding corona is accelerated to supersonic speed near the Sun. So scientists went looking for Alfven waves around the Sun.

The presence of Alfven waves in the solar wind was confirmed by the Mariner space probes in the 1960s and 1970s. Alfven waves were also detected outside the ecliptic, above the Sun's poles, by magnetometers on board Ulysses and the two Helios probes.

Scientists believe that Alfven waves exert pressure that can provide an extra boost to the heat-driven solar wind, accelerating it to a higher speed. But do Alfven waves exist close to the Sun?

In 2007, data collected from the Hinode space probe provided evidence that Alfven waves were being generated in the bottom of coronal holes on the Sun.

Fig. 8.3 Composite of EUV images taken on 23rd December 1996 by instruments on the SOHO spacecraft. The innermost image (*centre*) shows the corona at a temperature of about 2–2.5 million degrees. The electrically charged coronal gas is seen blowing away from the Sun just outside the inner *dark circle*, which marks the edge of one instrumental occulting disc. Three prominent coronal streamers can be seen (two at the *west* and one at the *east* limb). The field of view of this instrument encompasses 32 diameters of the Sun. To put this in perspective, the diameter of this image is 45 million kilometres at the distance of the Sun, or half of the diameter of the orbit of Mercury. The *centre* of the Milky Way is visible, as well as the dark interstellar dust rift, which stretches from the *south* to the *north*. This image also shows Comet SOHO-6 (elongated streak at about 7:30 h), one of several tens of sungrazing comets discovered so far by SOHO. It eventually plunged into the Sun (Credit: NASA, ESA/SOHO).

Table 8.1 Average solar wind parameters as measured by the Helios 1/2 probes

Parameter	Fast wind	Slow wind
Main source	Coronal holes	Equatorial streamers
Composition/density	Uniform	Variable
Magnetic field lines	Open	Closed
Proton speed	750 km/s	348 km/s
Proton temperature	280,000°C	55,000°C
Helium temperature	730,000°C	170,000°C

Further results indicate that the chromosphere is permeated with Alfven waves that are energetic enough to accelerate the solar wind.

The new findings will help modellers create improved Sun simulations. Many mysteries remain about the Sun's restless activities. Some scientists are focused on Alfven waves generated by the Sun's heat turbulence, but others are studying Alfven waves generated when the Sun's magnetic field lines stress and snap back together like invisible magnets.

Scientists still don't know which source of Alfven waves plays a more important role in the heating the Sun's atmosphere, but can use the latest findings as a stepping stone to better understanding.

The Heliosphere

Charged particles in the solar wind move away from the Sun in all directions, forming a huge three-dimensional bubble around the Sun called the heliosphere. Our knowledge of this sphere has come from spacecraft such as Ulysses orbiting the Sun. The solar wind gets thinner as it expands into a greater volume. Eventually it becomes too dispersed to exert any pressure. This occurs at a distance of about 100 astronomical units (AU) from the Sun (1 AU is the distance between Earth and the Sun). This distance is far beyond the orbit of the known outer planets.

The Voyager 1 and 2 spacecraft that are heading out of the solar system in different directions, crossed through the outermost boundary of the heliosphere in December 2004 and August 2007 respectively.

The solar wind does not travel in straight lines. Because the Sun rotates on its axis, the magnetic field around the Sun is pulled into a spiral shape within the plane of the Sun's equator. The shape has been confirmed by spacecraft orbiting the Sun. Charged particles in the solar wind are forced to follow these curved magnetic field lines. See Fig. 8.4.

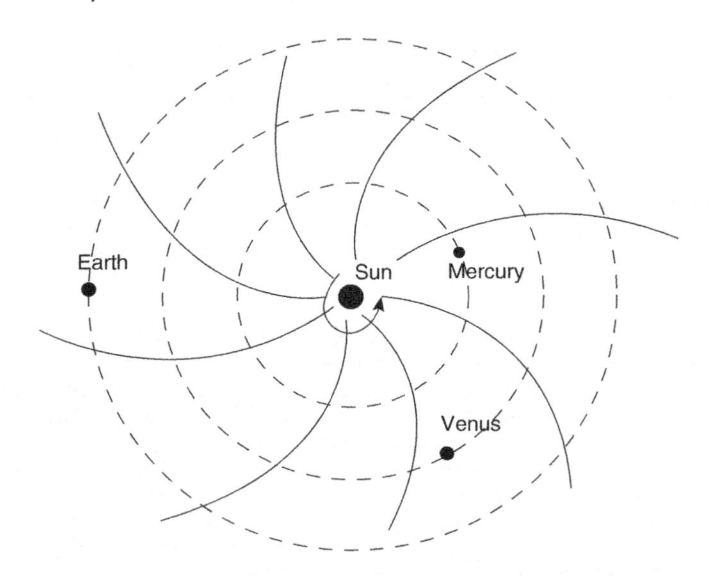

Fig. 8.4 The spiral shape of the Sun's magnetic field in interplanetary space as viewed from above the Sun's north pole. The approximate locations of the orbits of Mercury, Venus and Earth are shown as *circles*.

The Earth's Magnetic Field

The Earth has its own magnetic field called the **magnetosphere.** This is the region around Earth where the magnetic field has an effect on charged particles. The field lines surrounding the Earth are similar in shape to those of a simple bar magnet. The Earth's magnetic field is not uniformly distributed around the Earth. On the sunward side, it is compressed because of the solar wind, while on the other side it is elongated to around three earth radii.

The Earth's rotational axis is titled with respect to its magnetic axis. Thus true north does not correspond with magnetic north. See Fig. 8.5.

The magnetic field of our planet protects the Earth's surface from bombardment by energetic particles from space. Most of these particles come from the solar wind. Near the Earth, the particles in the solar wind move at speeds of roughly 400 km/s. Many of these particles are deflected by our planet's magnetic field but some get trapped in the Van Allen radiation belts (see Fig. 8.5). Instruments on board the Explorer 4, Pioneer 3 and Luna 1 space probes first mapped out the trapped radiation.

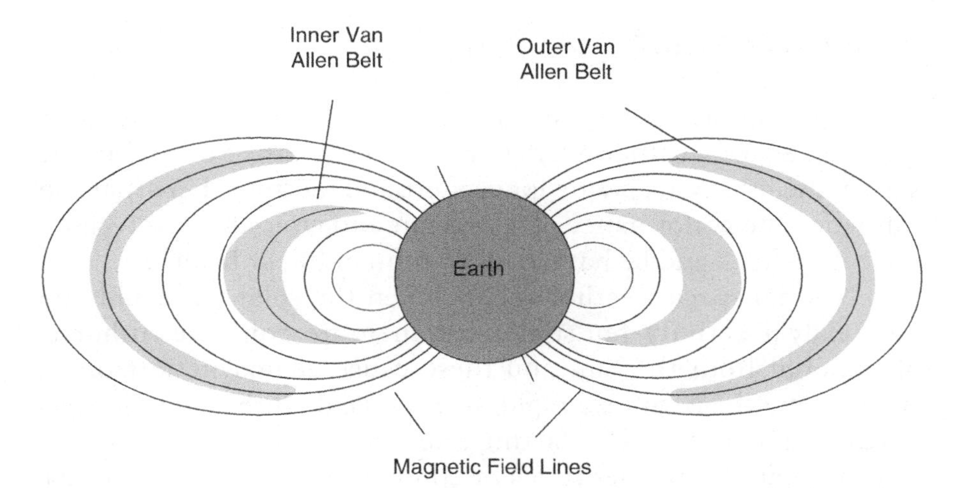

Fig. 8.5 The Earth's magnetic field and position of the Van Allen Radiation belts. Notice there is an inner and outer belt.

There is an inner and outer radiation belt. The particle population of the outer belt is varied, containing electrons and various ions. Most of the ions are in the form of energetic protons, but a certain percentage are alpha particles and oxygen ions, similar to those in the ionosphere but much more energetic. The inner belt contains high concentrations of energetic protons and electrons, trapped by the stronger magnetic fields in this region.

Sometimes the Van Allen belts become overloaded with radiation and the particles cascade down into the Earth's upper atmosphere. These high speed, charged particles collide with gases in the atmosphere causing them to fluoresce (i.e. emit light). The result is a beautiful, shimmering display of lights in the sky called the northern lights (aurora borealis) or the southern lights (aurora australis), depending on which hemisphere the phenomena is observed in. Auroras are seen best in polar regions but sometimes they can be seen in lower latitudes.

The radiation belts are a hazard for artificial satellites and are moderately dangerous for human beings, but are difficult and expensive to shield against. Solar cells, integrated circuits and sensors found in spacecraft can be damaged by this radiation. Electronics on satellites must be hardened against radiation to operate reliably. The Hubble Space Telescope has its sensors turned off when passing through regions of intense radiation.

Space Weather

In the past decade it has become fashionable to talk about space weather. Space weather refers to conditions on the Sun and in the solar wind, magnetosphere, ionosphere and thermosphere that can influence the performance of spacecraft and ground-based technological systems and be hazardous to human life or health.

Normal space weather occurs when the solar wind and Sun are steady in activity, but sometimes there are explosive outbursts of material from the Sun and these cause geomagnetic storms. A geomagnetic storm is a rapid, worldwide disturbance in Earth's magnetosphere, typically lasting a few hours. These storms usually originate from solar flares or coronal mass ejections (CMEs). They occur more often near the maximum of an 11-year solar activity cycle.

Energetic particles accelerated by solar flares or coronal mass ejections can cripple spacecraft and seriously endanger unprotected astronauts that venture into space. The storms can disrupt global communication systems and disable navigational and military satellites.

Energetic particles arriving at Earth from the Sun have the same ingredients as the steady solar wind, but they have much faster speeds and vastly greater energy. During solar flares, for example, protons and electrons can be accelerated to speeds more than 100 times that of the solar wind, with energies 10,000 times more than solar wind particles.

Potential solar outbursts can be forecast by monitoring changes in the magnetic fields in active-regions on the Sun. Scientists are unable to predict exactly when a solar flare or CME will begin, but once they occur it takes 1–2 days for the material to reach Earth, so alerts can be given. The SOHO spacecraft observes the emission of CMEs from the Sun and determines if they are headed towards Earth. SOHO measures incoming solar energetic particles and the solar wind and the data provides about 40 min warning before the particles strike Earth's magnetosphere. ACE uses real-time observations to provide short-term forecasts of shock-accelerated, high-energy protons at http://www.srl.caltech.edu/ACE/ASC/rtsw.html.

The launch of the NASA-ESA Solar-Terrestrial Relations Observatory (STEREO) added an additional space weather data stream that covers the region between the Sun and the Earth with stereoscopic imagery.

In addition, non-solar sources such as galactic cosmic rays, meteoroids and space debris can all be considered as altering space weather conditions near Earth. Note: Sunspots do not have much influence on geomagnetic storms.

Geomagnetic Storms and CME

The most major geomagnetic storms are induced by coronal mass ejections. CMEs are usually associated with flares, but sometimes no flare is observed. CMEs have more of an effect on space weather because they carry more material into a larger volume of interplanetary space. This increases the chances of an interaction with Earth's magnetosphere. A CME also produces a huge shock wave that continuously produces energetic particles as it moves through space. A flare on its own produces high-energy particles close to the Sun, but some of these particles escape into interplanetary space. See Fig. 8.6.

Particles produced by solar flares follow the interplanetary magnetic field lines, while particles from CMEs can cross magnetic field lines and accelerate particles all the way from the Sun to the Earth. Electrons accelerated in the outward propagating shock wave also produce Type II radio bursts that can be detected by radio telescopes on Earth.

When a CME travels out into interplanetary space, it is often called an **interplanetary coronal mass ejection** or ICME. Only about 10% of the CMEs observed with coronagraphs are detected as ICMEs – these mass ejections are generally faster and more energetic. Spacecraft investigating ICMEs have found that they temporarily block the flow of cosmic rays to our planet. As a coronal mass ejection moves into interplanetary space, it often takes the form of a magnetic cloud containing a mass of magnetic loops.

Fig. 8.6 This illustration shows a CME blasting off the Sun's surface in the direction of Earth. This *left* part of the picture contains an EIT 304 image superimposed on a LASCO C2 coronagraph. Two to four days later, the CME cloud is shown striking and beginning to be deflected around the Earth's magnetosphere. The *blue* paths emanating from the Earth's poles represent some of its magnetic field lines. These storms, which occur frequently, can disrupt communications and navigational equipment, damage satellites, and even cause *blackouts*. (Objects in the illustration are not drawn to scale) (Credit: NASA).

Did You Know?

One of the best-known examples of a space weather event is the collapse of the Hydro-Quebec power network on 13th March 1989 due to geomagnetically induced currents. This was started by a transformer failure, which led to a general blackout that lasted more than 9 h and affected six million people. The geomagnetic storm causing this event originated from a Coronal Mass Ejection that occurred on the Sun on 9th March 1989.

On 20th January 1994 a geomagnetic storm temporarily knocked out two Canadian communications satellites, Aniks E1 and E2 and the international communication satellite Intelsat K.

On 10th January 1997, a CME hit the Earth's magnetosphere and caused the loss of the AT&T Telstar 401 communication satellite that was worth $200 million.

Aeroplanes flying over Earth's polar regions are particularly sensitive to space weather. It is estimated to cost about $100,000 each time such a flight is diverted from a polar route. Nine airlines are currently operating polar routes.

No large space weather events have happened during a manned space mission. However, a large event did occur on 7th August 1972, midway between the Apollo 16 and Apollo 17 lunar missions. If a dose of particles ever hit an astronaut outside of Earth's protective magnetic field, during one of these missions, the effect would have been deadly or at least life threatening.

A large stream of solar energetic particles hit the Nozomi Mars Probe on 21st April 2002, causing large-scale failure. The mission, which was already about 3 years behind schedule, was eventually abandoned in December 2003.

One of the best-known ground-level consequences of space weather is geomagnetically induced current. These are damaging electrical currents that can generate power surges on transmission lines and current flow in pipelines and other conducting networks. Air and ship magnetic surveys can be severely affected by geomagnetic storms, resulting in data interpretation problems.

Auroras

Large geomagnetic storms often produce intense auroras in Earth's upper atmosphere. An **aurora** is a bright display of coloured light in the night sky, known popularly as the northern lights (aurora borealis) or southern lights (aurora australis). Auroras are produced when charged particles trapped in Earth's magnetic field collide with atoms in the upper atmosphere of Earth. At times of geomagnetic storms, there are extra charged particles in these regions and so auroras are prominent. Auroras are usually seen between 60° and 72° north and south latitudes, which places them in a circular zone just within the Arctic and Antarctic polar circles. The circular zone (also called an auroral oval) is about 4,000 km in diameter and centred on each magnetic pole. Auroras sometimes occur over the poles but such events are infrequent and often invisible to the naked eye. Brilliant auroras, associated with large magnetic storms, can extend down toward the Earth's equatorial regions.

Most auroras are seen as a diffuse glow or as a 'curtain' that roughly extends east west across the sky in the direction of the nearest pole to the observer. Each curtain consists of many parallel rays, each lined up with the local direction of the magnetic field lines. Sometimes aurora form 'quiet arcs' while at other times they continually change shape.

The active phase of an aurora will last 15–40 min and may recur in 2–3 h. Some auroral band features may last all night.

The visible light emission seen in auroras occurs around 90–240 km above the Earth's surface and is due to the excitation of oxygen and nitrogen atoms and molecules in the atmosphere by energetic electrons circulating in the magnetic field around the north and south poles.

Between the heights of 150 and 400 km above the Earth's surface, red aurora at 6,300 and 6,364 A wavelengths occur due to collisions of electrons with atoms of oxygen. Between 90 and 150 km the aurora is mainly green at 5,577 A because of denser layers of oxygen. Between 65 and 90 km the lower border may be tinged with red molecular emissions of nitrogen. See Fig. 8.7.

Blue auroral emissions are caused by ionised nitrogen molecules at 3,914 and 4,278 A. Red is also emitted from ionised nitrogen atoms at wavelengths of 6,611, 6,624, 6,696, 6,705, 6,789 and 6,875 A, when they emit an electron and return to the ground state. The colours

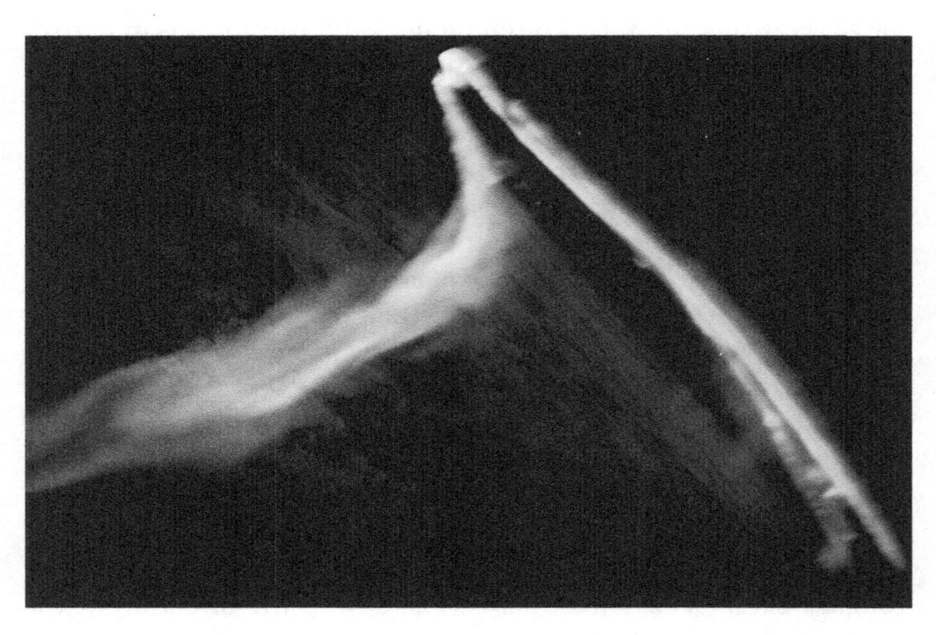

Fig. 8.7 An aurora captured by the crew of the International Space Station on 29th May 2010. The *green* light is due to excitation of neutral oxygen atoms in Earth's upper atmosphere by energetic electrons. Some stars can be seen through auroral displays. Imagine what this display would look like from the Earth's surface? (Credit: NASA/ISS and JSC).

emitted depend on altitude; at high altitude oxygen red dominates, followed by oxygen green and nitrogen blue/red. Green is the most common colour emitted. A pink or yellow aurora can occur by a mixture of red and green light (Fig. 8.8).

In polar regions, auroral activity peaks around times of sunspot maximum. The best auroral displays seem to occur near the equinoxes in autumn and spring, but can occur at any time. It is not known for certain why this is the case, but it is known that at these times the Earths magnetic field and the interplanetary magnetic field are aligned. Auroral displays at the poles can only be

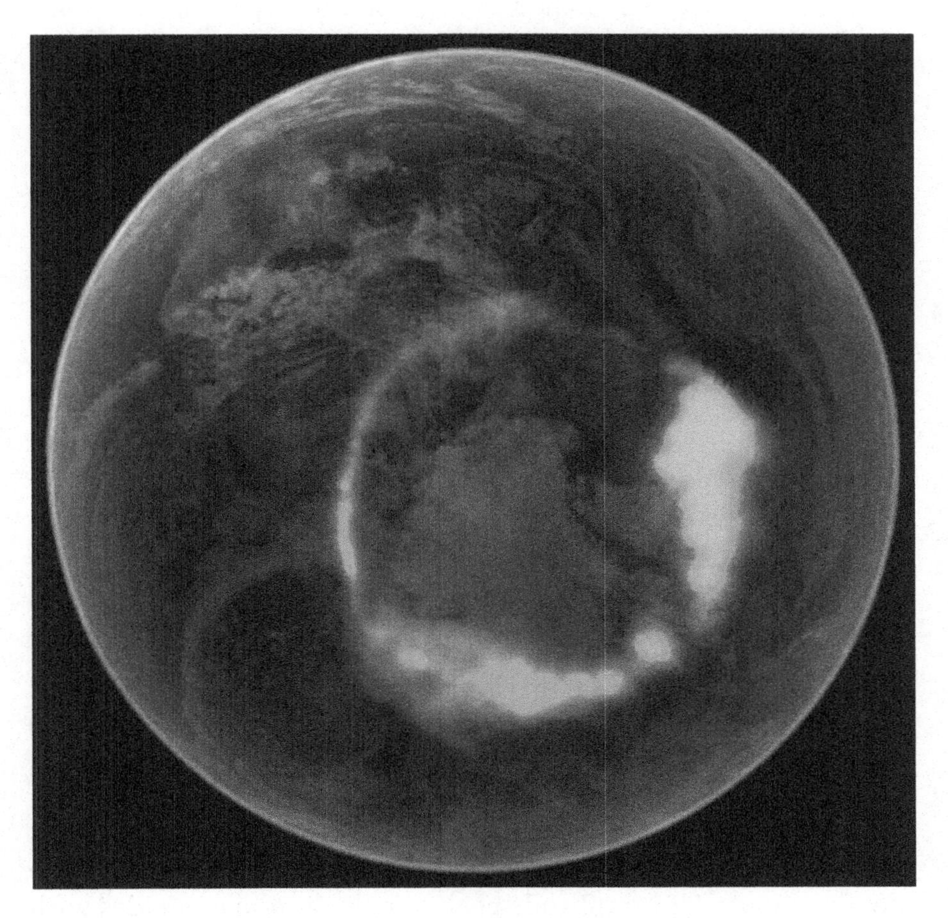

Fig. 8.8 Auroral Australis over Antarctica as captured by NASA's IMAGE satellite on 7th January 2005. The aurora was caused by a significant geomagnetic storm. The Image spacecraft was the first dedicated to imaging the Earth's magnetosphere (Credit: NASA/Image satellite).

seen during 6 months of the year when the pole is experiencing 24 h of night (for the south pole this is between March and September).

Displays of aurora borealis commonly occur in northern parts of Canada, where up to 200 displays a year are seen. Smaller numbers of auroras occur in northern Norway and Sweden. In the northern United States and southern Canada, auroras occur about a dozen times a year.

In the southern hemisphere, aurora australis is visible in Antarctica, but there are few people living there to see it. Sometimes aurora australis can be seen from southern parts of Australia, South Africa and South America.

A typical aurora requires an energy input of about 1,000 billion watts. During intense displays up to one million amperes of current can flow along an aurora with voltages round 50,000 V. This can induce currents to flow in unwanted places on Earth, such as Arctic pipelines and power grid networks.

Earth is not the only planet to have auroral storms around the poles. Both Jupiter and Saturn have magnetic fields stronger than Earth's and both have large radiation belts. Auroras have been observed on both planets by the Hubble Space telescope. These seem to be powered by the solar wind. Uranus and Neptune also have auroras around their poles.

Aurora and Radio Signals

Geomagnetic storms cause clouds of ions to form in the upper atmosphere of Earth at the base of auroras. These clouds can scatter or absorb radio waves even if an aurora is not visible. Ion clouds in the E layer of the atmosphere (112 km altitude) can absorb HF (high frequency) and scatter VHF (very high frequency) radio signals. This has the effect of increasing transmission distances in VHF from 80 km up to 1,600 km, especially when they lie on the same line of geomagnetic latitude. A radio operators 'horizon' expands with increasing strength of the aurora and the clouds of ions. The amplitude peak of radio aurora coincides with the peak of the break-up phase of visible aurora and the peak of magnetic disturbance. In Europe, radio aurora often peak in the afternoon and evening. The frequency of radio aurora also varies with the seasons (peaks in spring and autumn) and follows the sunspot cycle.

At times of severe geomagnetic storms, HF radio blackouts occur on the entire sunlit side of the Earth lasting for a number of hours – this results in no HF radio contact with mariners and en route aviators in this sector.

Amateur radio operators can monitor auroras using VHF equipment and a movable directional antenna. Directional results from two or more separate stations can be used to determine where a particular ion cloud is located.

Observing Aurora

Many amateur astronomers enjoy making observations of auroras and taking photographs of them. Obviously it is best if you live at the right latitude, that is, near the polar regions. When observing aurora, a sketch should be made (see Fig. 8.9) and the following factors recorded:

1. Location of observer: latitude and longitude need to be recorded.
2. Time of observations: record date and universal or local time.
3. Elevation: angular height from horizon to bottom of display and to the top.
4. Direction and span: compass bearing to each side of display.
5. Activity/Condition: note any movement as quiet (still), active or pulsating.
6. Appearance: glow, band, rays, curved arch, band, curtain etc. Uniform or irregular.
7. Brightness: weak, bright, very bright.
8. Colour: red, green, blue, pink, etc. Uniform or mixed.

Astronomical societies or clubs sometimes have auroral section that specialise in observations.

You need to be aware of times when auroral alerts are given (see last section in this chapter).

Time exposure photographs can be taken of auroras. Many amateurs have produced excellent photographs of auroras. Digital cameras have the advantage of giving instant results. A trial and error method is best to use to find out what exposures to use. For time exposures, a tripod and shutter release cable need to be

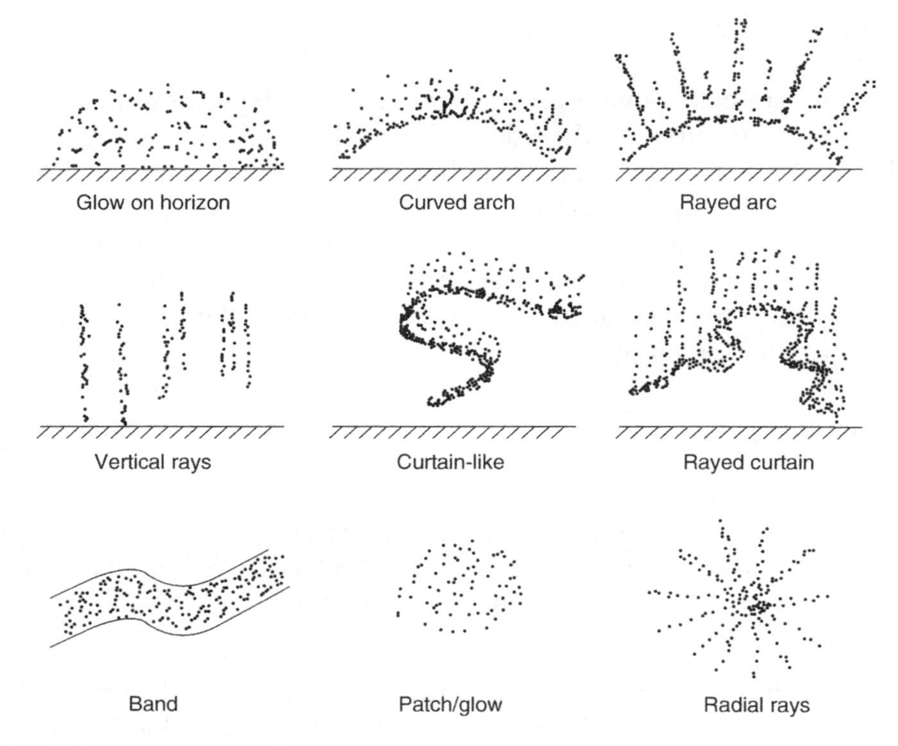

Fig. 8.9 Some types of aurora.

used. There are many samples of amateur photographs posted on websites under the heading of auroras.

Cosmic Rays

Cosmic rays are energetic particles that enter Earth's atmosphere from all directions in outer space. They consist mainly of hydrogen nuclei (protons), helium nuclei (alpha particles), electrons and positrons, and heavier nuclei. They can originate from galactic cosmic radiation but some come from CMEs on the Sun. Cosmic rays are observed indirectly by a device known as a neutron monitor. When cosmic ray particles enter the Earth's atmosphere they interact with the nuclei of the air molecules to produce secondary radiation. The neutrons predominate in this secondary radiation and the cosmic ray detector actually detects the secondary

Table 8.2 Composition of cosmic rays at top of Earth's atmosphere

Type of nucleus	Percentage abundance
Hydrogen nuclei (protons)	85
Helium nuclei (alpha particles)	12
Electrons and positrons	2
Heavier nuclei	1

neutrons. When cosmic ray particles enter the Earth's atmosphere they collide with molecules, mainly oxygen and nitrogen, to produce a cascade of lighter particles, a so-called air shower (Table 8.2).

The magnetic fields entrapped in and around coronal mass ejections exert a shielding effect on the galactic cosmic radiation that is detected by the neutron monitors. This causes a reduction in the count rate from the monitor, typically from about 3% to 20%. This reduction is a reliable indicator of a geomagnetic storm with warning times of up to 24 h or more.

Cosmic rays are classed as non-electromagnetic radiation. They travel with much more energy than solar protons. The outer layers of the Earth's atmosphere stop most cosmic rays so they do not reach the ground. Just as well, because high-speed ions can do a lot of damage to living organisms. However, people flying high in aeroplanes are subject to high levels of cosmic rays. Radiation doses from cosmic rays roughly increase exponentially with altitude. Repeated exposure can damage human DNA and possibly cause cancers.

The radiation from cosmic rays composes a large part of natural background radiation on the surface of Earth. It is far more intense outside the Earth's atmosphere and magnetic field, and because it is very difficult to shield against, is expected to have a major impact on the design of spacecraft that can safely transport humans in interplanetary space.

Tracking Solar Storms

Scientists use space probes and ground-based telescopes to track solar outbursts from their beginning on the Sun, to their passage through space to Earth. Instruments aboard SOHO and Hinode, for example, monitor the varying magnetic fields in the photosphere, as well as the extreme ultraviolet or x-ray emissions from solar flares. Coronagraphs aboard SOHO and the two STEREO probes detect CMEs. ACE, Wind and Ulysses can follow the progress of solar flare electrons and CME driven shocks as they move through interplanetary space.

Closer to Earth, the GOES (Geostationary Operational Environmental Satellite) probe monitors flaring x-ray radiation and energetic particles. GOES orbits Earth once every 24 h in a geostationary orbit at a height of 35,790 km. GOES 1 was launched in 1975 and several others have been launched since. GOES images the Earth for short term terrestrial weather forecasting and storm tracking and it also monitors the space environment – including variations in the Sun's soft x-ray flux received at the spacecraft. GOES also classifies soft x-ray flares as A, B, C, M or X, from weakest to strongest, according to peak detected flux. Each class has a peak flux ten times greater than the preceding one (flux is measured in Watts per square metre).

By combining all this data, scientists get a good picture of what is heading towards Earth and can estimate the effect.

Space Weather Websites

There are a number of websites that provide information on the current space weather. A number of these contain similar data. You are advised to view each site as detailed below and use one that suits your own requirements. Some are easy to use than others, and some provide information of little value to amateurs.

Spaceweather.com

The website http://www.spaceweather.com is a useful one for amateur solar astronomers. It contains news and information about the Sun-Earth environment. On the left hand side of the main page is a list of current conditions on the Sun, including solar wind speed (km/s) and density (protons/cm^3), a list of recent x-ray solar flares and an image of the Sun's current active regions. Lower down on the left hand side is the current Sunspot Number (this is the 'Boulder number' as explained in Chap. 3); radio flux details; an image of the auroral oval over Earth's poles; the Interplanetary magnetic field strength; an image showing any coronal holes; NOAA forecast showing the probability of flares and geomagnetic storms (active, minor storm, severe storm).

On the main page there is an archive of Aurora Photos going back several years and a list of recent and upcoming Earth-asteroid encounters. There are also links to other 'space weather' websites.

The SOHO Website

The SOHO website http://sohowww.nascom.nasa.gov contains satellite information about various aspects of space weather on the right hand side of main page. Click on the 'Space weather' box and nine more boxes appear (See Fig. 8.10). Note; NOAA = US National Oceanic and Atmospheric Administration, SWPC = Space Weather Prediction Centre.

1. If you click on the first box headed (NOAA/SWPC) the latest GOES satellite solar x-ray image comes up together with a 3-day solar-geophysical forecast. There is also a graph of solar x-ray flux and Proton flux. The GOES 12 through 15 spacecraft each carry a sophisticated solar x-ray Imager to monitor the Sun's x-rays for the early detection of solar flares, coronal mass ejections, and other phenomena that impact the geo-space environment.
2. The second box contains a 3-day graph of the GOES x-ray flux. The GOES x-ray Flux plot contains 5-min averages of solar x-ray output in the 1–8 Å and 0.5–4.0 Å bandpasses. SWPC x-ray alerts are also issued at the M5 and X1 levels.
3. The third box contains three graphs showing soft x-ray output of the Sun over 3 days from SOHO/SEM/EUV.
4. The fourth box contains a coloured dial showing the geomagnetic storm level based on ACE solar wind data. Blue = low, Green = medium, Yellow = high, Red = extreme.

5/6. The fifth and sixth boxes show auroral activity from the NOAA-POES satellite. There are two maps showing the extent of the auroral oval over each pole (north and south).

7/8. The seventh and eighth boxes show graphs of the proton levels of the Sun.

Fig. 8.10 Nine boxes on SOHO space weather website.

NOAA SWPC 1.	GOES X-ray flux 2.	X-ray flux SOHO/SEM 3.
Geomagnetic ACE 4.	Avroral activity north pole 5.	Avroral activity south pole 6.
Proton flux 7.	Proton levels 8.	ACE solar wind data 9.

9. The ninth box shows ACE solar wind real-time data (speed and pressure). If you click on this box, two more boxes come up, one showing the NOAA activity scale for storms, the other showing a graph of the current solar cycle.

The most useful boxes for amateur solar astronomers are 1, 4, and 9 (Fig. 8.10).

If you go back to the main SOHO page, there is also a long box headed 'solar wind' on the right hand side showing current solar wind speed and density.

Space Weather Prediction Centre

The Space Weather Prediction Centre (SWPC) has a website at www.swpc.noaa.gov. The site lists current space weather conditions as well as links to data from ACE, Stereo, GOES and POES satellites. On the right hand side is a useful section headed: NOAA Scales Activity – this section uses a scale from G1 (minor) to G5 (extreme) to rate current Geomagnetic storms, S1 to S5 to rate solar radiation storms and R1 to R5 to rate Radio blackouts.

There is also a K-index (introduced by Julius Bartels in 1938/9) that quantifies disturbances in the horizontal component of the

Earth's magnetic field. The index ranges from 1 (calm) to 9 (severe geomagnetic storm). The K-index however varies from observatory to observatory as it depends on latitude and other factors. The Kp index (logarithmic) is derived by calculating a weighted average of K numbers from a network of geomagnetic observatories. The Kp scale is a reasonable way to summarize the global level of geomagnetic activity, but it has not always been easy for those affected by the space environment to understand its significance. In general if the Kp index is above 5, there is a real chance of aurora occurring. (Note: there is also a box for 'Estimated Kp' on the SOHO website).

The Ap index gives daily averages on a linear scale. An Ap index of 0–10 is considered to represent a quiet field, 10–20 represents a minor storm, 20–50 a storm, and over 50, a major storm (Fig. 8.11).

The NOAA G-scale was designed to correspond to the significance of effects of geomagnetic storms (see Table 8.3).

Fig. 8.11 A quiet day on the Sun can still be pretty complicated if you were a weather forecaster there. This image of the northeastern solar limb was taken by TRACE on 17th October 2000 in the 171 Å bandpass, showing emission from gas at approximately one million degrees. In the *lower-left* corner is Active Region 9199, with a double filament (*F1*) reaching across it; matter is flowing to the *left*, following the *red* curves. There are at least five other filaments (*F*) in this image, plus a dark, cool arcade of loops (*A*), and a short-lived jet (*J*) (Credit: NASA/TRACE).

Table 8.3 NOAA space weather scale used to measure geomagnetic storms

NOAA G scale	Descriptor	Corresponding Kp value	No. of storm days per 11 year sunspot cycle
G5	Extreme	9	4
G4	Severe	8	100
G3	Strong	7	200
G2	Moderate	6	600
G1	Minor	5	1,700

The Australian Space Weather Agency

The Australian Space Weather Agency has a web site at http://www.ips.gov.au/Space_Weather.

The information is presented in an easy to follow format. Parts are useful for amateur solar observers. The data is divided into five sections:

1. Solar conditions: solar wind speed, x-ray flux, x-ray flares, latest Culgoora spectrograph, and latest Culgoora H-alpha image.
2. Physical conditions: geomagnetic warning, GEOSTAT alert, Geomagnetic alert, Aurora alert, K-index, pc3-Index, AusDst-Index.
3. HF propagation conditions: HF Radio communications warning, Current HF fadeout, HF fadeout warning, and Polar cap absorption.
4. Ionosphere conditions: Australasian and World maps are given.
5. Total Electron Current (TEC) conditions: Maps given for Australasia and World.

To get information about each of the parameters listed above (for example, K-index, pc3-Index) go to the section headed 'Geophysical help page'.

The Australian Space Weather Agency gets some of its information from the Culgoora Solar Observatory. This observatory is located 25 km west of the town of Narrabri, in northwest New South Wales, Australia. The observatory conducts continuous optical and radio observations of the Sun every day of the year.

Observing instrumentation includes:

- A 12 cm solar telescope fitted with a hydrogen-alpha filter, used to observe solar flares and other phenomena
- A 30 cm heliostat, used to observe sunspot evolution

- A solar radio spectrograph which sweeps through a frequency range of 18–1,800 MHz every 3 s, used to monitor solar radio bursts

Regular reports and forecasts of solar activity are transmitted to the Australian Space Forecast Centre in Sydney and are disseminated to similar organizations internationally. Particularly significant solar outbursts are reported to a wide range of interested parties around the world within minutes of their occurrence.

Web Notes

ACE uses real-time observations to provide short-term forecasts of shock-accelerated, high-energy protons at: http://www.srl.caltech.edu/ACE/ASC/rtsw.html.

The website http://www.spaceweather.com is a useful one for amateur solar astronomers.

The SOHO website http://sohowww.nascom.nasa.gov contains satellite information about various aspects of Space weather.

The website of the Space Weather Prediction Centre is http://www.swpc.noaa.gov.

The Australian Space Weather Agency has a very good web site: http://www.ips.gov.au/Space_Weather

The Culgoora Solar Observatory, Australia has a web site at: http://www.ips.gov.au/Solar/2/1

9. The Sun and Earth's Climate

The Intergovernmental Panel on Climate Change (IPCC) claim that our planet is undergoing a change in climate. Part of their reasoning is based on the fact that the average global surface temperature has increased by approximately 0.3–0.6°C over the last century. This is the largest increase in surface temperature in the last 1,000 years and scientists are predicting an even greater increase during this century. Such increases are thought to be responsible for changes in sea levels, precipitation and weather patterns. Many scientists think that the Sun is the main cause of climate change while others think it is due to human activities. This chapter will focus on the effect the Sun and human activities have on our climate as well as other natural influences.

Climate and Weather

For centuries, the annual seasons and weather patterns on Earth have been driven by the sunlight that reaches the lower atmosphere and surface. When most people talk about the weather, they are really talking about things like rain, storms, sunshine, temperature, clouds and wind. All these things refer to the condition of the air that surrounds us – that is, the weather.

Climate refers to the sum of all weather events in an area over a long period of time. Thus climate and weather is not the same thing. The weather may change from day to day. One day's weather may be stormy, wet and cool, while the next day's weather may be sunny, dry, and warmer. To determine the climate in a given area, scientists need to study the weather conditions over many years.

Scientists who study the climate are called climatologists. They describe the climate in a particular area in terms of average

J. Wilkinson, *New Eyes on the Sun*, Astronomers' Universe,
DOI 10.1007/978-3-642-22839-1_9, © Springer-Verlag Berlin Heidelberg 2012

temperature and rainfall. They also consider variations that occur between the different seasons of the year. Climates differ from area to area on the Earth's surface because of differences in:

1. Latitude,
2. The availability of moisture,
3. Land and water temperatures, and
4. The topography of the land.

The Atmosphere

All Earth's weather occurs in the atmosphere. The atmosphere consists of a mixture of the gases nitrogen, oxygen and carbon dioxide, as well as other minor gases, water vapour and particles of dust and smoke. The atmosphere extends from the ground to an altitude of about 160 km. Above this altitude there is very little air and it gets thinner the higher you go.

Nearly all the weather occurs in the lower part of the atmosphere in a layer called the **troposphere**. The troposphere extends from the surface to a height of about 16 km. Air and clouds continually move about in this layer, and the temperature decreases with increasing height (from 20°C at sea level to −55° at 16 km). Weather conditions in the troposphere depend on four things – air temperature, air pressure, wind strength and moisture content or humidity (Fig. 9.1).

The atmosphere is heated by the Sun. About 47% of the Sun's rays that enter the atmosphere reach the surface to warm the seas and land. About 19% of the Sun's rays are absorbed by the atmosphere and warm it, while 34% is reflected back into space. Because the atmosphere traps a large amount of heat, it acts like a "greenhouse". Greenhouses contain walls and a roof of glass that allow heat from the Sun in but prevent the reradiated heat from escaping. See Fig. 9.2.

Naturally occurring "greenhouse gases" in our atmosphere have a mean warming effect of about 33°C. The major greenhouse gases are water vapour, which causes about 36–70% of the greenhouse effect; carbon dioxide (CO_2), which causes 9–26%; methane (CH_4), which causes 4–9%; and ozone (O_3), which causes 3–7%. Clouds also affect the radiation balance, but they are composed of

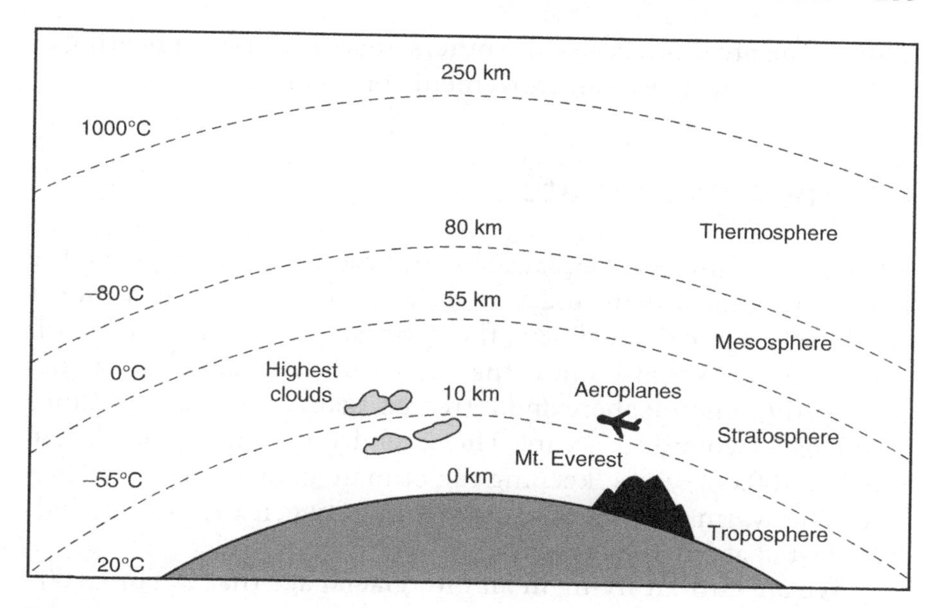

Fig. 9.1 Layers of the Earth's atmosphere.

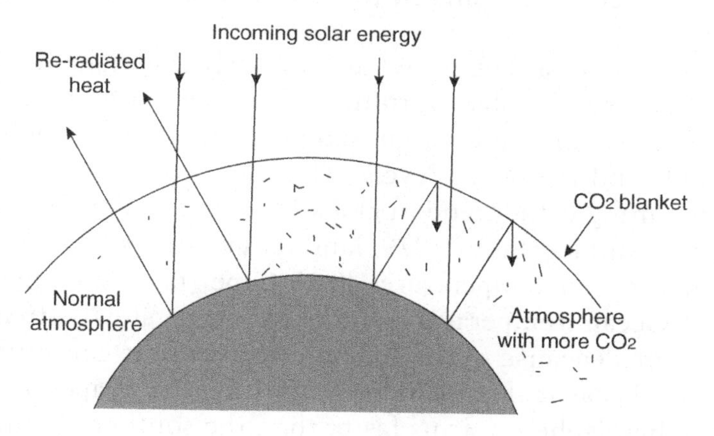

Fig. 9.2 The "Greenhouse effect" occurs when the layer of gases in the atmosphere traps the Sun's heat. Increased amounts of CO_2 gas in the lower atmosphere traps more solar radiation, thus raising the temperature of the air. As a result the Earth's surface becomes warmer.

liquid water or ice and so have different effects on radiation from water vapour.

Because of clouds and latitude differences, not all parts of the Earth's surface receive the same amount of heat from the Sun.

Thus some areas are hot while others are cold. Uneven heating of the atmosphere causes air movements or winds.

Changes in Climate

Changes in climate take place slowly over a number of years. We know that our climate today is different from the warm, moist conditions that existed during the Permian Period, about 200 million years ago. We also know that ice sheets have advanced across the Northern hemisphere and retreated again more than 20 times in the past two million years. The extended ice sheets have lasted roughly 100,000 years, keeping the climate cold and the sea level low. The warm periods that existed in between the cold periods have lasted about 10,000 years.

We are current living in an interglacial age that began 10,000 years ago when the Earth was about 5°C warmer and wetter. The ice sheets melted and shrank to there current positions and sea levels rose.

Many scientists believe we are currently experiencing climate change because of global warming. There have been increases in global average air and ocean temperatures, faster melting of snowfields and rising sea levels. The most common measure of global warming is the trend in globally averaged temperature near the Earth's surface. Since 1979, land temperatures have increased more than ocean temperatures (0.25°C per decade compared to 0.13°C). Ocean temperatures increase more slowly than land temperatures because of the higher effective heat capacity of the oceans and because the ocean loses more heat by evaporation. The northern hemisphere warms faster than the southern hemisphere because it has more land and extensive areas of seasonal snow and sea-ice. More greenhouse gases are emitted in the northern than southern hemisphere but this does not contribute to the difference in warming because the major greenhouse gases spread from one hemisphere to the other.

Climatologists believe there are a number of causes of climate change. One cause might be a variation in the amount of energy received from the Sun. Another cause might be dust and soot thrown into the air as a result of volcanic eruptions or burning.

A third cause might be an increase in the amount of carbon dioxide and other gases released into the atmosphere by human activities. Lets look more closely at each of these causes.

Varying Solar Energy

Climate change might be caused by variations in the amount of energy we receive from the Sun. Such variations can arise from any or all of three events that are part of a natural cycle. Each event affects the intensity of the seasons on Earth:

1. A short periodic wobble in the Earth's rotational axis that is repeated every 23,000 years.
2. A longer periodic variation of the Earth's axial tilt occurs every 41,000 years. It is currently 23.5° but varies between 21.5° and 24.5°. The greater the tilt the hotter the summers and the colder the winters.
3. The third and longest cycle is due to a slow periodic change in the shape of the Earth's orbit every 100,000 years. As the orbit becomes more elongated, the Earth's distance from the Sun varies more during the year and this influences the seasons.

These natural changes need to be taken into consideration when discussing the causes of climate change. Because these changes occur over long periods of time, there is little accurate data available to understand the effects.

Volcanic Activity

We know that in Earth's past there has been a lot of volcanic activity. When a volcano erupts, it throws huge amounts of dust into the atmosphere. The dust may stay in the air for many years, scattering the Sun's rays and reducing the amount of sunlight that reaches the ground. Thus a volcanic eruption may have short term cooling effects. A similar effect may be caused by an asteroid or large meteoroid crashing into the Earth.

When Mount Pinatubo erupted in the Philippines on 15th June 1991, an estimated 20 million tonnes of sulfur dioxide and ash particles blasted more than 20 km high into the atmosphere. The eruption caused widespread destruction and loss of human life. Gases and solids injected into the stratosphere circled the globe for 3 weeks. Volcanic eruptions of this magnitude can affect global climate, reducing the amount of solar radiation reaching the Earth's surface, lowering temperatures in the troposphere, and changing atmospheric circulation patterns. The extent to which this occurs is an ongoing debate.

Burning fires can also put dust and soot into the atmosphere. Soot may cool or warm the surface, depending on whether it is airborne or deposited. Airborne soot directly absorbs solar radiation, which heats the atmosphere and cools the surface. In areas with high soot production, such as rural India and parts of Asia, as much as 50% of surface warming due to greenhouse gases may be masked by atmospheric brown clouds.

Greenhouse Gas Levels

Scientists have learnt about past climate from studying microscopic air bubbles trapped inside glacial ice. Analyses of glacial ice cores reveal roughly 100,000 years periodicity of the ice ages over the past 420,000 years. The results indicate transitions from cold to warm periods are accompanied by increases in the concentration of the greenhouse gases, carbon dioxide, methane and nitrous oxide. Temperatures seem to go up when the levels of these gases increase and temperatures go down when they drop (see Fig. 9.3). The ice cores can also be used to determine solar activity during past centuries.

In the past 100 years the Earth has on average warmed up by more than half a degree. The increase in global temperature seems to coincide with large increases in the amount of carbon dioxide and other heat-trapping gases in the atmosphere. These gases have come from human activity such as burning coal, gas and oil for energy. The situation is becoming worse as the worlds human population increases and countries become more industrialised. A number of scientists believe that without remedial action, the amount of carbon dioxide in the atmosphere will soon be double that of last century. This will cause temperatures near the surface to rise to uncomfortable levels.

In 2010 the concentration of CO_2 in the atmosphere was 390 parts per million by volume (ppmv). For the past 10 years it has

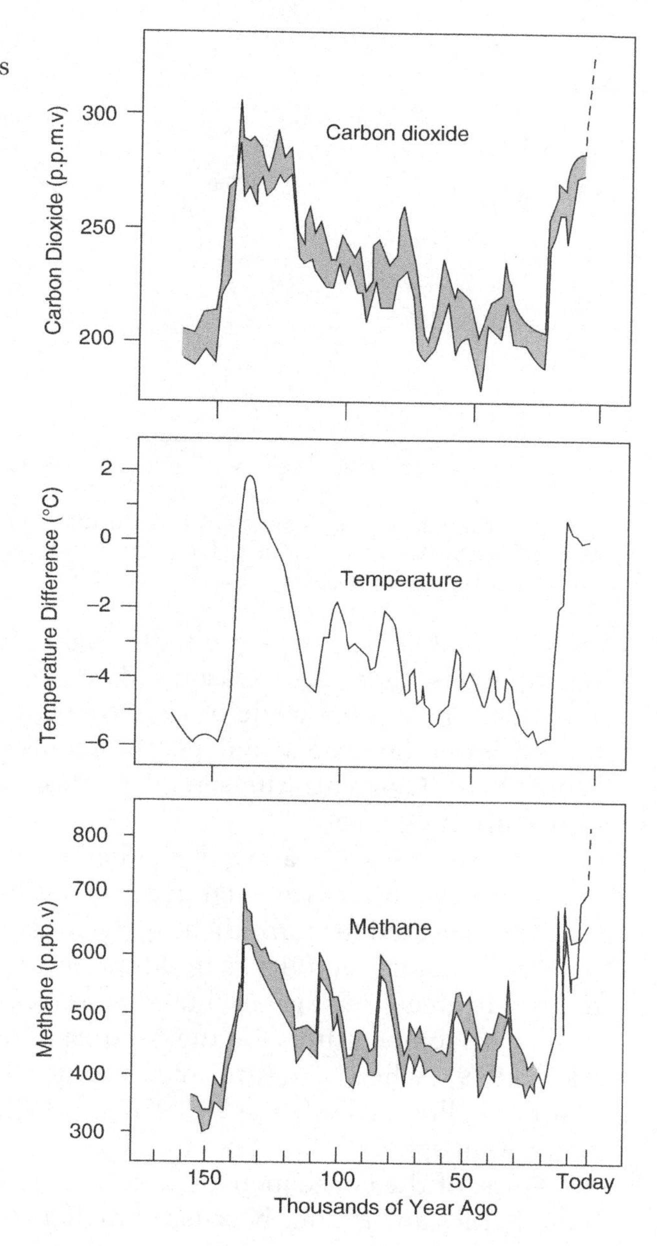

Fig. 9.3 Ice-core data indicates that changes in atmospheric temperatures over Antarctica closely follow variations in the concentration of the gases carbon dioxide and methane for the past 160,000 years. Note: ppmv = part per million by volume.

been rising by 1.6 ppmv each year. At this rate, a concentration of 450 ppmv will be reached in 37 years and this equates to a global temperature increase of 2.6°C. This time is very short with respect to the timescale of evolution (see Fig. 9.4). There is an annual CO_2

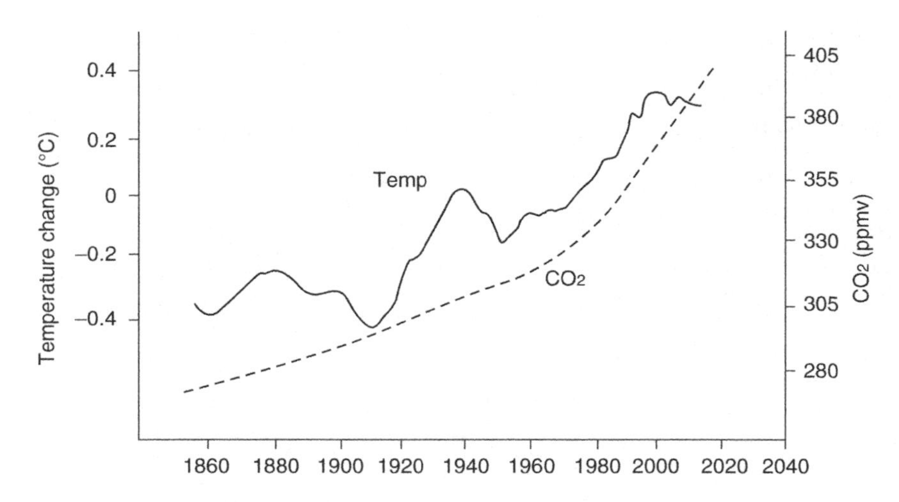

Fig. 9.4 Graph showing changes in the amount of carbon dioxide (CO_2) in the Earth's atmosphere (smoothed) together and the temperature change or anomaly (average 12 month values) since 1860.

fluctuation of about 3–9 ppmv that roughly follows the Northern Hemisphere's growing season. The Northern Hemisphere dominates the annual cycle of CO_2 concentration because it has much greater land area and plant biomass than the Southern Hemisphere. Concentrations tend to peak in May and reach a minimum in October.

Greenhouse gases and solar radiation affect temperatures in different ways. Increased solar activity and increased greenhouse gases are expected to warm the troposphere. In the stratosphere an increase in solar activity should produce warming, while an increase in greenhouse gases should produce cooling. Observations show that temperatures in the stratosphere have been cooling since 1979, when satellite measurements became available. Weather balloon data between 1958 and 1979 also shows cooling in the stratosphere.

Some of the consequences of global warming are: flooding of coastal cities and islands because of higher sea levels; stronger and wetter hurricanes; reduced fresh water supplies and more forest fires; more droughts inside continents; more heat waves. Current international agreements to reduce the production of carbon dioxide and other heat-trapping gases is unlikely to solve the problem unless large polluting countries like the USA and China reduce emissions as well.

The Ozone Layer

Ozone (O_3) is a form of oxygen that is present in the Earth's atmosphere in small amounts. Most ozone is found in the atmosphere at altitude 23–30 km in a layer called the ozone layer. This layer is important to life as it shields the Earth from 95% to 99% of the Sun's ultraviolet rays. The release of chemical gases (like halocarbons) into the atmosphere by humans is destroying the ozone layer, allowing more ultraviolet radiation to reach the surface. Such radiations can be detrimental to human health.

The amount of ozone in the atmosphere is also altered by variations in the flow of both UV radiation and energetic particles from the Sun. We know, for example, that the enhanced UV radiation that pours outward from the Sun at times of high solar activity increases the amount of ozone in the atmosphere. At times of minima in the 11-year cycle, less ozone is found. Since ozone in turn affects the lower atmosphere and biosphere, it provides another possible connection between solar variability and the climate system.

Ozone depletion occurs in many places in the Earth's ozone layer, most severely in the polar regions. In 1986, scientists reported a "hole" in the ozone layer over Antarctica. The hole is a region of ozone depletion in the stratosphere that happens at the beginning of the southern hemisphere spring (August – October). Satellite images provide us with daily images of ozone levels over the Antarctic region (Fig. 9.5).

The Antarctic ozone hole was once regarded as one of the biggest environmental threats, but the discovery of a previously undiscovered feedback (reported in 2010) shows that it has instead helped to shield this region from carbon-induced warming over the past two decades. High-speed winds and a lot of sea spray in the area beneath the hole have led to the formation of brighter summertime clouds, which reflect more of the Sun's heat away from the surface to the extent that warming from rising carbon emissions has effectively been cancelled out in this region.

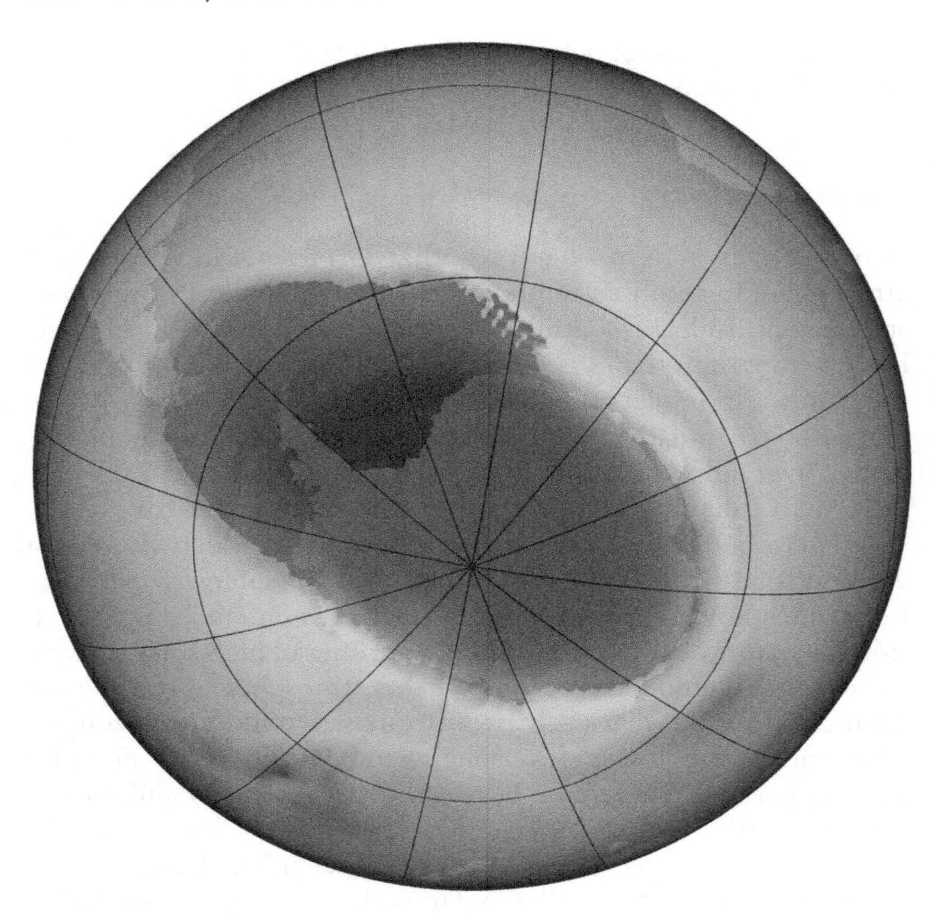

Fig. 9.5 The ozone hole is the region over Antarctica with total ozone of 220 Dobson Units or lower. This image shows the ozone hole (*purple*) on 22nd September 2004. The data was acquired by the Ozone Monitoring Instrument on NASA's Aura satellite. Aura was launched on 15th July 2004 in a polar orbit around Earth at an altitude of 705 km (Image courtesy the Scientific Visualization Studio, NASA/GSFC).

Significant depletion also occurs in the Arctic ozone layer during the late winter and spring period (January-April). However, the maximum depletion is generally less severe than that observed in the Antarctic, with no large and recurrent ozone hole taking place in the Arctic.

NOAA (National Office of Oceanic and Atmospheric research) researchers regularly measure ozone depleting gases in the lower and upper atmosphere and attempt to account for observed changes. As a result of international regulations, ozone-depleting gases are being replaced in human activities with "ozone-friendly" gases that have much reduced potential to deplete ozone. NOAA researchers are also measuring these "substitute" gases as they accumulate in the atmosphere. Observing changes in both old and new gases emitted into the atmosphere allows researchers to improve our understanding of the fate of these gases after release and thereby improve our ability to predict future ozone changes.

Did You Know?

During the last 400 years, the Earth has experienced the extremes of solar-driven climate change.

First a very cold, dry climate, then a much warmer and wetter climate and now the cold dry climate again. The period of approximately 1600–1700 AD is known as the "The Little Ice Age". This cold climate was the result of a period of very low sunspot activity called the "The Maunder Minimum". Famines, plagues and droughts caused the collapse of many cultures reducing the world's population by one third. During this period the average world temperature cooled by about 1.5°C compared to the peaks in the preceding 4,000 years. After the Dalton Minimum (1800–1830 AD), the average solar radiation rapidly increased again until peaking in the mid 1970s. This was the highest average peak for 8,000 years. This peak produced a rapid rise in the world's average temperature and also resulted in high rainfall averages. Since the mid 1970s the average levels of solar radiation have been rapidly declining again. The effects can now be seen very clearly. Firstly, the average rainfall declined in parallel with reducing levels of solar radiation. Secondly, after 1998, the global temperature peaks also commenced declining in line with the 20-year time lag required for thermal inertia loss from the world's oceans. This declining trend in global temperature peaks since 1998 has altered the prediction that the continuing rise in carbon dioxide levels would produce a corresponding rise in world temperatures. Actually in recent years the global average has declined and is now about 0.15°C lower than the peak year.

Sunspot experts from both Russia and the US are forecasting that the present 30-year decline in solar radiation will continue well into the future. A leading Russian expert says the world will cool and produce another Little Ice Age period by about 2042.

Climate and the Sunspot Cycle

Scientists have tried to link changes in climate with variations in the number of sunspots on the Sun. The total amount of sunlight received by the Earth varies only slightly with the 11-year sunspot cycle. However the amount of the Sun's ultraviolet and x-ray radiation, which is absorbed by the Earth's atmosphere, varies by larger amounts during the solar activity cycle. At periods of solar maximum the Earth is exposed to higher levels of electromagnetic radiation as well as greater amounts of energetic solar particles but fewer cosmic rays. These factors may alter conditions in our atmosphere and contribute to climate changes.

Throughout most of the twentieth century most scientists and climatologists thought the Sun was shining at a constant rate and any variations in the amount of sunlight reaching the ground were attributed to variable absorption and scattering in the atmosphere. In the early 1980s, instruments on board the Solar Maximum Mission satellite were able to measure total solar radiation levels to within 0.01% or one part in 10,000. At this level of accuracy, the scientists found the Sun's output is almost always changing by amounts up to a few tenths of a percent. This variation is linked to the concentrated magnetism in sunspots that produce reductions in the Sun's output of up to 0.3% over a few days. Bright patches called faculae cause increased luminosity. Over the short term, sunspot blocking appears to have a larger effect than faculae brightening, but in the long term faculae are dominant because they last longer than sunspots and cover a larger area of the Sun's disc.

Improved instruments on satellites over the past few decades have enabled scientists to monitor the Sun's output with greater accuracy. The change in output is about 0.07% from cycle minimum to maximum. At times of high solar activity, the Sun emits much more invisible radiation – especially ultraviolet and x-rays. The atmosphere of Earth absorbs this radiation and the upper atmosphere heats up. When solar activity is at minimum, the upper atmosphere absorbs less and cools down.

Precise solar irradiance measurements obtained during the past three decades show that variations in total solar irradiance (TSI) occur over time scales from minutes to 11-year solar cycles and longer. Climate models including sensitivity to solar forcing estimate a global climate change of up to 0.2°C due to solar variations over the last 150 years (Fig. 9.6).

To determine long-term changes in the Sun's output, which may have time scales extending much longer than the 11-year solar cycle, the TSI climate record requires either very good absolute accuracy or very good instrument stability and continuous measurements. To date, no TSI instrument has achieved the

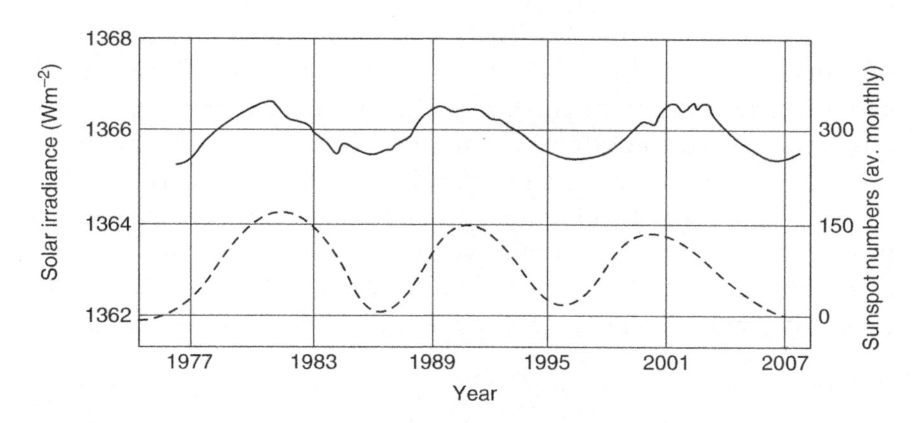

Fig. 9.6 Annual variations in the Sun's total irradiance (output) as measured by satellite instruments at the top of Earth's atmosphere between 1977 and 2007. The output varies with the 11-year sunspot cycle. Sunspot maximums occurred in 1979, 1990 and 2001, while minimums occurred in 1987, 1997, and 2008. The average value for solar irradiance is about 1,366 W m^{-2}. Dashed line indicates sunspot numbers.

necessary absolute accuracy, and the TSI record relies on measurement continuity from overlapping spacecraft instruments.

In 1991, Eigel Friss-Christensen and Knud Lassen published a paper showing a close correlation between the length of the sunspot cycle and temperature variations over the past 130 years. They found average air temperatures over land in the Northern Hemisphere had varied by about 0.2°C in synchronism with the sunspot cycle length during this period. However, cloud cover, drought, rainfall, tropical monsoons and forest fires showed variable correlations with solar activity.

In 1997, Henrik Svensmark and Eigil Friis-Christensen found that during the previous 11-year sunspot cycle, the global cloud cover decreased with increasing solar activity and a lower intensity of cosmic rays entering the atmosphere.

In October 2010, the science journal "Nature" reported that a group of scientists claimed that a decline in solar activity could actually lead to a warmer Earth not a cooler one as previously suspected. The study used data collected by the NASA-sponsored SORCE satellite, which allowed the scientists to fully break down the Sun's light output into its different wavelength components for the first time. The SORCE data revealed that even though total solar irradiance declined from 2004 to 2007, the amount of visible light output from the Sun had actually increased. Unlike other wavelengths of light, visible light cuts through different atmospheric layers to warm the surface of the Earth directly, creating an overall warming effect. If this were true, it would mean that climate scientists have been overestimating the contributions of the Sun on climate change and underestimating the effect of human activity. Only 3 years worth of SORCE data are currently available, and scientists will have to monitor solar activity for at least a complete solar activity cycle to test their ideas.

Did You Know?

The ancient civilizations like the Greeks and the builders of Stonehenge believed that the planetary cycles and lunar cycles played an important part in shaping the climate of that part of the world. A number of present-day weather forecasters also use these cycles to predict long-term weather. The understanding is that these two cosmic cycles can each enhance or diminish average rainfall over decadal periods resulting in extended flood and drought cycles.

Lunar cycles are varied and complex. We know that the Moon goes through a cycle of phases each month but it also affects the rise and fall of ocean tides and causes tides in the air. The Moon also changes in declination – from being high in the sky to being low in the sky. The maximum and minimum declination reached by the Moon varies because the Moon's orbital plane precesses (wobbles) through an 18.6-year cycle. These cycles affect the weather on Earth. The 18.6-year lunar cycle is believed to be responsible for lifts in annual rainfall up to 150% at the peak of the cycle. At the bottom of the cycle (approximately every 9.3 years) there are drought years with only 50% of average rainfall.

The planetary cycle is determined by the positions of planets in the Solar System, especially Jupiter and Saturn, in relation to Earth. For example, when Saturn and Jupiter are on opposite sides of the Sun, rainfall is enhanced. This occurs on average once every 19.86 years. Ten years later when Saturn and Jupiter are together on the same side of the Sun, reduced rainfall for several years usually occurs. The magnitude of rain-enhancement depends on which planets are close to the Earth at a particular time and how the sea-surface temperature anomalies (La Nina and El Nino) have developed.

For about 60 years of the 297-year full combined cycle, the lunar cycle and the planetary cycles are closely in-phase, and the extremes of climate (flood then drought) are most likely to occur 9–10 years apart (the period 2011–2012 is an in-phase time).

Interactions between these cycles is more complicated than discussed here, however, long-term weather forecasters who use these cycles claim that the effects of these cycles are much more direct and powerful than any alleged effects of human-driven CO_2 climate change.

The Overall Picture

It should be clear that there are a number of different factors that might be causing Earth's climate to change. The impacts of these factors are mixed together, and further confused by imperfect knowledge of how each of them has changed, and uncertainties in how climate itself has varied. There are many pieces of contradictory data, especially with regards to graphical data of climate changes. For example, solar irradiance data needs to be measured in the atmosphere – this has only been accurately done since 1978. Graphical data showing values long before this time are unreliable and often use ground based data (which is different).

What is also unclear is how the Earth itself will respond to climate altering changes. One way to monitor the changes is to keep measuring the mean-surface and atmospheric temperature of Earth. The sensitivity of climate to solar radiation changes is not well known. A conservative estimate is that a 0.1% change in solar total radiation will bring about a temperature response of 0.06–0.2°C, providing the change persists long enough for the climate system to adjust. This could take 10–100 years.

Changes in visible and infrared solar radiation alter the surface temperature by simple heating. Other types of radiation have less direct effects on climate; for example, the UV radiation that is emitted by the Sun at times of high solar activity increases the amount of ozone in the stratosphere, while at times of low solar activity less ozone is found. These solar variations make it hard to determine the effects of human-activities on the ozone layer.

An analysis of climate altering factors for the period from 1850 to 1990, has shown that some have caused warming (e.g. increasing radiation from Sun) while others have caused cooling (e.g. more ozone). The net result is an addition of 1.2 W per square metre. The Sun accounts for about a quarter of the net amount. Most of the change is attributed to factors associated with human activities. The most likely cause of climate change in the period since about 1850 is the growing concentration of greenhouse gases: in particular carbon dioxide (CO_2), methane (CH_4), nitrous oxide (N_2O), and the commercially-made substances called halocarbons. Climate simulations using only greenhouse gas changes predict a warming that exceeds the 0.5°C that is documented in the instrumental record of the past 140 years. To reconcile the difference between the observed and the predicted values, either the models are wrong or other factors must be properly factored in.

If we assume that the climate is equally sensitive to each of these causes, the net increase of 1.2 W should have brought about an increase in global mean temperature of 0.3–1.1°C. The documented rise of about 0.5°C in the same period falls at the low end of this range. It may be premature to make such a comparison, however, since it is uncertain when all of the warming would be felt, given the lag times of up to a century that are imposed on the climate system by the thermal inertia of the oceans.

Without doubt the Sun plays a critical part in the Earth's climate system, moreover, both the Sun's output and the climate, change continually over all time scales. One and a half decades of continuous monitoring of direct solar radiation have provided long-needed information, but this short period of time is but a tiny one in the life of the Sun, and an inadequate sample of the full range of its possible behaviour. Detecting and confirming larger-amplitude, longer-period cycles in solar radiation will require reliable continuous solar monitoring by satellites, well into the next century.

The rapid global warming since 1970 is several times larger than that expected from any known or suspected effects of the Sun, and may already indicate the growing influence of atmospheric greenhouse gases on the Earth's climate. However we must allow for the fact that solar changes could potentially alter the anticipated effects of carbon dioxide and other greenhouse gases on the surface temperature of the Earth.

Web Notes

For information on global warming try: http://solar-center.stanford.edu/sun-on-earth/glob-warm.html or http://www.global-greenhouse-warming.com/solar-irradiance-measurements.html

For an article on solar activity and climate change see: http://www.sciencedaily.com/releases/2009/03/090328163643.htm

For an article on sunspots, volcanic eruptions and climate change see: http://unisci.com/stories/20022/0613022.htm

The Sun and climate: http://www.gcrio.org/CONSEQUENCES/winter96/sunclimate.html

For an article on volcanoes and climate change see: http://earthobservatory.nasa.gov/Features/Volcano/ and http://www.geo.mtu.edu/volcanoes/vc.../o_sc_volcano_climate.html

For an article of the ozone layer see: http://www.oar.noaa.gov/climate/t_ozonelayer.html and http://www.theozonehole.com/climate.htm and http://ozonewatch.gsfc.nasa.gov/

To find out more about long range weather forecasts (particularly in Australia) based on the lunar and planetary cycles see: http://www.TheLongView.com.au

10. The Sun and Stars

The Sun is a star around which the Earth and other planets of the Solar System orbit. Like any star, the Sun is a glowing ball of hot ionised gas or plasma. The light and heat the Sun generates comes from nuclear reactions inside the Sun's core. These reactions mainly involve the fusion of hydrogen into helium. As a result the Sun generates a tremendous outward pressure that is balanced by the inward pull of the Sun's gravity, holding the star in a near spherical shape. This state is known as 'hydrostatic equilibrium' and stars spend most of their lives in this state.

When we look up into the night sky we see stars, many of which are like the Sun. Not all stars are the same; they differ in size, brightness, colour, composition and life span. Today we know the distances to stars with a high degree of precision. We have even found that many stars have planets orbiting them, just like our Sun has.

In this chapter we shall examine our Sun as a star and compare it with other stars.

Distance to Stars

Up until 500 years ago, it was generally assumed that the Earth was the centre of the universe and that the stars were fixed to a giant celestial sphere that surrounded the Earth. Today we know that the Earth is not the centre of the universe and that the distance to stars is huge compared to distances in our Solar System. Astronomers measure the distance to stars in units of light years. One light year (ly) is the distance light travels in 1 year. Light travels at 300,000 km/s so in 1 year it travels 9,500,000,000,000 km or 9.5×10^{12} km.

J. Wilkinson, *New Eyes on the Sun*, Astronomers' Universe,
DOI 10.1007/978-3-642-22839-1_10, © Springer-Verlag Berlin Heidelberg 2012

The nearest star to the Sun is Proxima Centauri at a distance of 4.2 light years (or 40 trillion kilometres). It takes light just over 4 years to get to Earth from this star. Proxima Centauri is a faint star and you need a telescope to see it. It is close to the bright star Alpha Centauri (4.3 ly). Alpha Centauri is the nearest star (apart from the Sun) we can see from Earth with our unaided eyes. It is in the constellation Centaurus and is one of the 'pointers' to the Southern Cross (visible in the southern hemisphere).

Astronomers measure the distance to stars using different methods – for close stars they use a method called **stellar parallax**. As the Earth moves from one side of its orbit around the Sun to the other, a nearby star's apparent position against the background of stars appears to shift. The parallax angle is half the angle by which a star shifts, measured in arc seconds (see Fig. 10.1). If the parallax

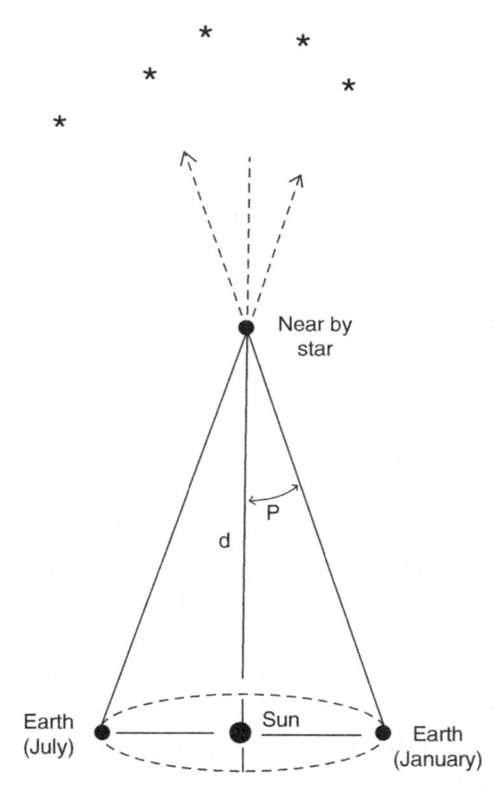

Fig. 10.1 Method of using parallax to determine the distance to stars close to Earth. The angle p is equal to the angular size of the radius of the Earth's orbit as seen from the star. The smaller the parallax the larger the distance to the star.

angle is p, the distance d to the star in parsecs is given by the equation

$$d = 1/p$$

Proxima Centauri has a parallax angle of 0.77 arcsec, and so its distance is 1/0.77, or approximately 1.3 parsec. One parsec (pc) = 3.26 light years (ly). Sometimes the parsec is used as the unit for distances to stars and galaxies.

Stellar parallax angles are extremely small and so the method is suitable for use only on close stars (up to about 100 pc). Astronomers have used satellites (with wider orbits than that of Earth), to measure distances to stars up to 500 pc away. The method of measuring distance to stars beyond 100 light-years involves the use of Cepheid variable stars. These stars change in brightness over time, which allows astronomers to figure out their true brightness. Comparing the apparent brightness of a star to its true brightness allows an astronomer to calculate the distance to the star. This method was discovered by American astronomer Henrietta Leavitt in 1912 and was used in the early part of the century to find distances to many globular clusters.

Brightness of Stars

Stars have been classified for centuries according to how bright they are. The brightest stars were originally called first magnitude stars, and their brightness designated as +1. Those about half as bright were called second magnitude stars, designated +2, and so forth, to sixth magnitude stars, the dimmest ones seen by the unaided eye. Astronomers today use a similar but more accurate scale of apparent magnitude. A first magnitude star is 2.512 times brighter than a second magnitude star, which is 2.512 times brighter than a third magnitude star, etc. Thus a first magnitude star is about 100 times brighter than a sixth magnitude star. Some stars are brighter than +1 magnitude, and astronomers have had to use negative numbers for the magnitude of the very brightest stars. For example, Sirius, the brightest stars in the night sky, has an apparent magnitude of −1.5. The dimmest stars visible through a small telescope have magnitude +10. Most stars visible to the unaided eye have magnitudes between +1 and +6.

Apparent magnitudes do not tell us about the actual energy emitted by the stars. A star that looks dim in the night sky might be brilliant but appears dim because it is very far away. To determine the energy emitted by a star, astronomers need to know how bright they really are compared to each other; that is, how bright they would be if they were all at the same distance from Earth. The **absolute magnitude** of a star tells us how bright a star would be if it were at a distance of 10 pc from Earth. If the Sun were moved to this distance it would have absolute magnitude +4.8 and Proxima Centauri +15.5. Absolute magnitudes range from −10 for the brightest stars to +17 for the dimmest. Astronomers can calculate absolute magnitudes from apparent magnitudes using a mathematical formula.

The energy emitted by a star is directly related to its brightness. **Luminosity** is the total energy emitted by a star from its surface each second. The luminosity of stars is expressed in multiples of the Sun's luminosity, which is equal to 3.90×10^{26} W. The brightest stars have luminosities 10^6 (one million) times that of the Sun. The dimmest stars have luminosities of 10^{-5} times that of the Sun. The Sun is an average star, with both its luminosity and absolute magnitude in the middle range for all stars. Table 10.1 lists the 15 brightest stars as seen in the night sky in order of brightness.

Temperature of Stars

Stars have different colours. For example, in the constellation Orion, Betelgeuse appears red and Rigel appears blue. A star's colour tells us a lot about its surface temperature. Red stars have temperatures between 2,000° and 3,000°, orange stars 3,000–5,000°, yellow stars 5,000–8,000°, white 8,000–12,000°, blue 12,000–35,000°. Temperatures can be in degrees Celsius (°C) or Kelvin units (K) – there is not much difference at high temperatures. To measure the colour of a star accurately, astronomers use a photometer or stellar spectrometer.

Table 10.1 The 15 brightest stars (in order of brightness)

Star name	Constellation	Magnitude		Distance	
		Apparent	Absolute	Parsecs	Light years
Sirius	Canis Major	−1.44	1.5	2.63	8.58
Canopus	Carina	−0.74	−5.6	96	310
Alpha Centauri	Centaurus	−0.28	4.1	1.34	4.37
Arcturus	Bootes	−0.05	−0.3	11.3	36.7
Vega	Lyra	0.03	0.6	7.76	25.3
Capella	Auriga	0.08	−0.5	12.9	42.2
Rigel	Orion	0.15	−6.8	240	780
Procyon	Canis Minor	0.38	2.7	3.50	11.4
Achernar	Eridanus	0.45	−2.8	44.1	144
Betelgeuse	Orion	0.50	−5.2	131	430
Hadar	Centaurus	0.61	−5.4	161	525
Acrux	Crux	0.74	−4.2	98	320
Altair	Aquila	0.75	2.2	5.13	16.7
Aldebaren	Taurus	0.87	−0.6	20.0	65
Antares	Scorpius	0.96	−5.1	150	490

The Sun is an average star – its surface temperature is around 5,600°C and its colour is yellow.

Size of Stars

The size of a star is linked to the amount of energy it emits and its temperature. Astronomers can determine the size of a star from its absolute magnitude, temperature and distance. The size of large stars can also be measured from interferometry. Another method involves analysing the light curves and radial velocities of eclipsing variables; this allows the dimensions of two stars to be calculated.

There are many stars which are similar in size to the Sun, and many which are smaller and some that are much larger. Stars whose diameters and luminosities exceed the Sun's by a factor of 100 or more are termed **supergiants**; those which are tens of times larger than the Sun are termed **giants**; those comparable with the Sun are **dwarfs** – the smallest of which are much fainter than the Sun are sometimes called **red dwarfs**; the very smallest are **white dwarfs**. See Table 10.2 and Fig. 10.2. One fairly well known example of a red dwarf star is Proxima Centauri – the closest star to Earth. This star has about 12% the mass of the Sun, and about 14%

Table 10.2 Stars like the Sun (Sun's mass = 1)

Star	Distance (light years)	Relative mass (Sun = 1)
Alpha Centauri A	4.4	1.10
Alpha Centauri B	4.4	0.90
Epsilon Eridani	10.5	0.80
Tau Ceti	12.2	0.82
Sigma Draconis	18.2	0.82
Delta Pavonis	19.2	0.98
82 Eridani	20.9	0.91
Beta Hydrii	21.3	1.23
Zeta Tucanae	23.3	0.90
Beta Canum Venaticorum	27.35	1.08
Gliese 67	41.42	0.97
Gliese 853	44.35	1.0
18 Scorpii	45.82	1.01
51 Pegasi	50.21	1.06

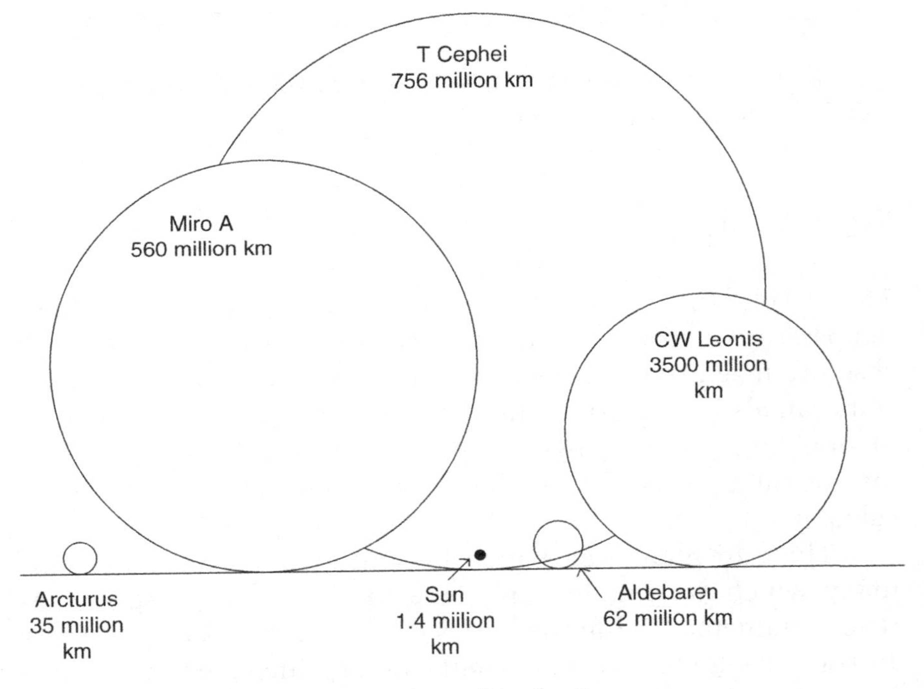

Fig. 10.2 Sizes of some stars compared to the Sun.

the size of the Sun – about 200,000 km across, which is only a little larger than Jupiter. An example of a star larger than our Sun is the blue supergiant Rigel in the constellation Orion. Rigel is 17 times the mass of the Sun and it puts out 66,000 times as much energy.

There are many stars about the same size as the Sun with a similar mass. For comparison purposes, the masses of stars are often given in terms of the Sun's mass (which is one unit). See Table 10.2.

Spectral Classification of Stars

The main method used by astronomers to determine the temperature of a star is by the spectroscopic analysis of its light. This method is also used to classify stars and to determine their chemical composition.

Spectral types are indicated by the letters OBAFGKM. The hottest stars, O stars, are blue and have surface temperatures over 35,000°C; their spectra is dominated by ionised helium and silicon. Only a few hundred of these stars are known and only three can be seen with the unaided eye: Regor, Zeta Puppis, Zeta Orionis. B stars are also blue and are known as the neutral helium stars since their temperature is not high enough to ionise helium. Examples of B stars include: Achernar, Rigel, and Bellatrix. Type A stars are blue-white in colour and have strong lines of hydrogen and ionised metals in their spectra. Examples of type A stars include: Sirius, Vega, and Deneb. Type F stars are white and show strong metallic lines (e.g. Ca II) with weak hydrogen lines. The brightest examples of F type stars are Canopus, Procyon and Polaris. G type stars are yellow and are known as Sun like stars – they are characterised by ionised calcium, iron and other metallic lines. Examples of G type stars include Capella, Alpha Centauri, and the Sun itself. K type stars are orange in colour and exhibit metallic lines of ionised calcium and iron together with neutral calcium and strong molecular bands. Examples of K type stars include Arcturus and Aldebaren. M stars are red and the coolest stars with a surface temperatures around 3,000°C. Their spectra are dominated by titanium oxide, and neutral metals. M stars are so cool that many of their atoms stick together to form molecules. Giant M type stars include Betelgeuse, Antares, and Mira. There are many M stars that are dwarfs, including Barnard's star and the two components of Kruger 60.

Astronomers have found it useful to subdivide the OBAFGKM temperature sequence further by adding an integer from 0 (hottest) to 9 (coolest). Thus an F8 star is hotter than an F9 star, which is hotter than a G0 star, which is hotter than a G1 star, and so on. The Sun's spectrum is dominated by singly ionised metals (especially Fe II and Ca II) and is a G2 star.

Around 1905 the Danish astronomer Ejnar Hertzsprung found that a pattern emerged when he plotted the absolute magnitudes of stars against their colour indices. Almost a decade later the American astronomer Henry Norris Russell independently discovered this regularity using spectral types instead of colour indices. Graphs showing the absolute magnitude of stars (or Luminosity) plotted against spectral type (or temperature) are called **Hertzsprung-Russell diagrams or HR diagrams**. In such a diagram, bright stars are near the top of the diagram, dim stars are near the bottom. Hot stars are on the left side of the graph, while cool stars are near the right side. See Fig. 10.3.

When stars are plotted on the HR diagram, they fall into distinct regions. The band stretching diagonally across the diagram represents most of the stars we see in the night sky. This band is

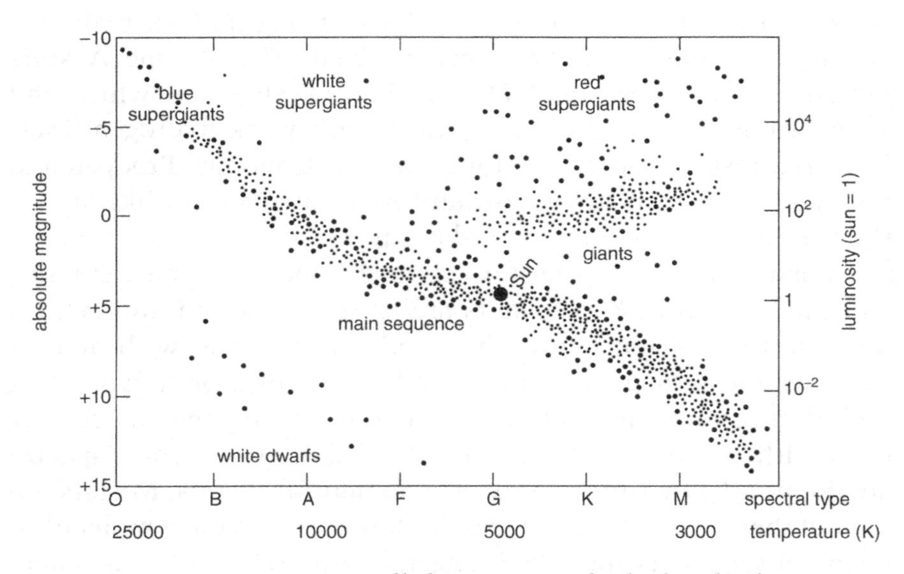

Fig. 10.3 A Hertzsprung-Russell diagram on which the absolute magnitude of stars is plotted against their Spectral type. The Sun's position is marked.

called the **Main Sequence** and 90% of stars fall into this band. Main sequence stars are fusing hydrogen in their cores. The hotter, brighter stars are more massive than the cooler dimmer stars. The Sun is a main sequence star with spectral class G2 and absolute magnitude +4.8. The number of main sequence stars decreases with increasing surface temperature. Above and to the right of the main sequence is a group of stars called the **giants**. These stars are typically 10–100 times as large as the Sun and have surface temperatures between 3,000° and 6,000°C. The coolest members of this group are called red giants and they appear red in our night sky. Aldebaren in the constellation Taurus and Arcturus in Bootes are examples of red giants that you can see with your unaided eye at night. Red giants are undergoing nuclear fusion by converting helium into carbon in their core, with hydrogen fusion in their outer layers.

To the left of the giants is a group called the **supergiants**. Supergiants are larger and brighter than the giants. They have multiple nuclear fusions in their outer layers with elements up to iron forming in their core. Betelgeuse in Orion and Antares in Scorpio are two examples of supergiant stars that we can see at night. Together, giants and supergiants comprise only one percent of the stars in our neighbourhood. The remaining 9% of stars fall into the lower and bottom part of the HR diagram. These stars are hot, dim and tiny compared to the Sun. They are called **white dwarfs** and are mainly the remnants of stars. White dwarfs are mainly made of carbon, and fusion has ceased in their cores – they are simply cooling down. They are so tiny that we need a telescope to see them.

Stars are also classified according to their luminosity and this is often included with spectral type. The supergiants are luminosity Class I, the bright giants are Class II, giants Class III, subgiants are Class IV, and the main sequence stars are Class V. White dwarfs are not assigned a luminosity class because they are just the remnants of stars; no fusion is taking place inside them. The Sun is classed as G2 V star while Aldebaren is a K5 III star. Sirius, the brightest star in our night sky has spectral type A1 V. Details of the spectral type of the 15 closest stars to the Sun are given in Table 10.3.

Table 10.3 Details of the 15 nearest stars

Star name	Constellation	Magnitude		Spectral type	Distance in light years
		Apparent	Absolute		
Sun	–	–26.72	4.85	G2 V	–
Proxima Centauri	Centaurus	11.09	15.53	M5.5 V	4.23
Alpha Centauri	Centaurus	0.01	4.38	G2 V	4.37
Barnard's star	Ophiuchus	9.53	13.22	M4.0 V	5.96
Wolf 359	Leo	13.44	16.55	M6.0 V	7.78
Lalande 21185	Ursa Major	7.47	10.44	M2.0 V	8.29
Sirius	Canis Major	–1.44	1.46	A1 V	8.58
Luyten 726-8	Cetus	12.54	15.40	M5.5 V	8.73
Ross 154	Sagittarius	10.43	13.07	M3.5 V	9.68
Ross 248	Andromeda	12.29	14.79	M5.5 V	10.32
Epsilon Eridani	Eridanus	3.73	6.19	K2 V	10.52
Lacaille 9352	Piscis Austrinus	7.34	9.75	M1.5 V	10.74
Ross 128	Virgo	11.13	13.51	M4.0 V	10.92
Luyten 789-6	Aquarius	13.33	15.64	M5.0 V	11.27
Procyon	Canis Minor	0.38	2.66	F5 IV	11.40
61 Cygni	Cygnus	5.21	7.49	K5.0 V	11.40

Star Formation and Stellar Evolution

The main thing to learn from the HR diagram is that there are different types of stars. These different kinds of stars represent different stages of stellar evolution. To truly appreciate the HR diagram we need to understand the life cycles of stars; that is, how they are born, what happens as they mature, and how they die. Astronomers have come to understand stellar evolution by observing stars with different temperatures, brightness, and chemical composition.

Stars form from the condensation of giant clouds of molecular gas and dust. Often these clouds have several hundred dense cores and many 'protostars' form inside them. Star formation begins when gravitational attraction causes the material in protostars to coalesce. As the protostar contracts, its gases begin to glow from the heat generated by the compression of the material it contains. When the contraction slows down, the protostar becomes a pre-main sequence star. When the pre-main sequence star's core temperature becomes high enough to begin hydrogen fusion, it becomes a main sequence star. The more massive a pre-main sequence star is, the quicker it begins hydrogen fusion in its core.

For example, a star of five solar masses takes 100,000 years to begin fusion, while a star of only one solar mass takes tens of millions of years to do the same. Pre-main sequence stars begin their life on the right hand side of the HR diagram. As they grow they move towards the main sequence band – the path taken is dependent on their mass. Stars spend most of their life on the main sequence. Main sequence stars are those fusing hydrogen into helium in their cores (Fig. 10.4).

The most massive main sequence stars are the most luminous, while the least massive stars are the least luminous. The more massive a star is, the faster it goes through its main sequence phase. O and B stars consume all their core hydrogen in a few million years. Stars of very low mass take hundreds of billions of years to convert their cores from hydrogen into helium. Our Sun will spend about ten billion years on the main sequence. See Table 10.4.

A star with a mass less than 0.1 of the mass of the Sun will continue to shrink but will never get hot enough for nuclear fusion to begin. It will fade away to form a small red star before turning cold and dying.

Fig. 10.4 A star forming region of the Large Magellanic Cloud. The image was taken using the Hubble Space Telescope (Credit: NASA/ESA/Hubble).

Table 10.4 Lifetimes of main sequence stars

Mass relative to Sun (Sun = 1)	Surface temperature (K)	Luminosity relative to Sun (Sun = 1)	Time on main sequence (millions yrs)
25	35,000	80,000	3
15	30,000	10,000	15
3	11,000	60	500
1.5	7,000	5	3,000
1.0	6,000	1	10,000
0.75	5,000	0.5	15,000
0.50	4,000	0.03	200,000

A star with a mass equal to that of the Sun initially glows very brightly. After this initial period, the star settles down to a long stable middle-life period of about ten billion years. Our Sun is now at about midlife and has another five billion years to go before it runs out of fuel. As these stars mature much of their hydrogen is used up, and the surface temperature decreases. When little hydrogen remains, the star's core shrinks, the outer layers expand and cool, and the star forms a red giant. After this, the gases in the outer regions drift into space and the remaining gases collapse into a small, very dense object known as a white dwarf. White dwarfs cool down and eventually fade away. Betelgeuse, the red star in Orion, is a red giant and was probably once similar in size to the Sun.

Stars about five times the mass of the Sun have a much shorter but spectacular life. These large stars live for only about one million years. The extra mass in these stars creates enormous gravitational forces in the core of the star. The nuclear reactions use fuel very rapidly, creating very hot bright stars that glow blue in the night sky. When the fuel runs out there is a tremendous outburst of energy and the stars brightness increases a million times. Such an event is called a **supernova**. Supernovas do not occur very often, since most stars in the universe are of medium size. The material blown out from a supernova forms a **nebula**. Some nebulas emit their own light and glow like stars. The remaining part of the star is pulled inwards by very strong gravitational forces to form an incredibly dense star less than 20 km in diameter. The original protons and electrons in the core are squeezed together so much that they fuse to form neutrons. This type of star is called a **neutron star**. Neutron stars do not glow very brightly, but they send out pulsating radio signals. Rapidly rotating neutron stars are called **pulsars**.

When stars about ten times the mass of the Sun become red supergiants they eventually explode as a supernova. After the supernova explosion, the core collapses on itself, and unlike a smaller star, these massive stars keep on collapsing. The gravitational force is so strong it will not even allow light to escape. What remains of the massive star is a hole in space which astronomers call a **black hole**. Radio astronomers have detected x-rays coming from invisible regions in space where black holes are thought to exist. Other black holes have also been detected in the centre of galaxies. See Figs. 10.5 and 10.6).

Did You Know?

One of the first white dwarf stars discovered was Sirius B, the companion star to Sirius – the brightest star visible in the night sky. Sirius B took a while to discover because it is about 10,000 times fainter than Sirius. The binary nature of Sirius was first deduced in 1844 by the German astronomer Friedrich Bessel, who noticed that the star was moving back and forth, as if orbited by an unseen object. The companion, Sirius B was first seen in 1862.

The Sirius double star system is about 8.7 light years away from Earth and is the fifth closest star system known.

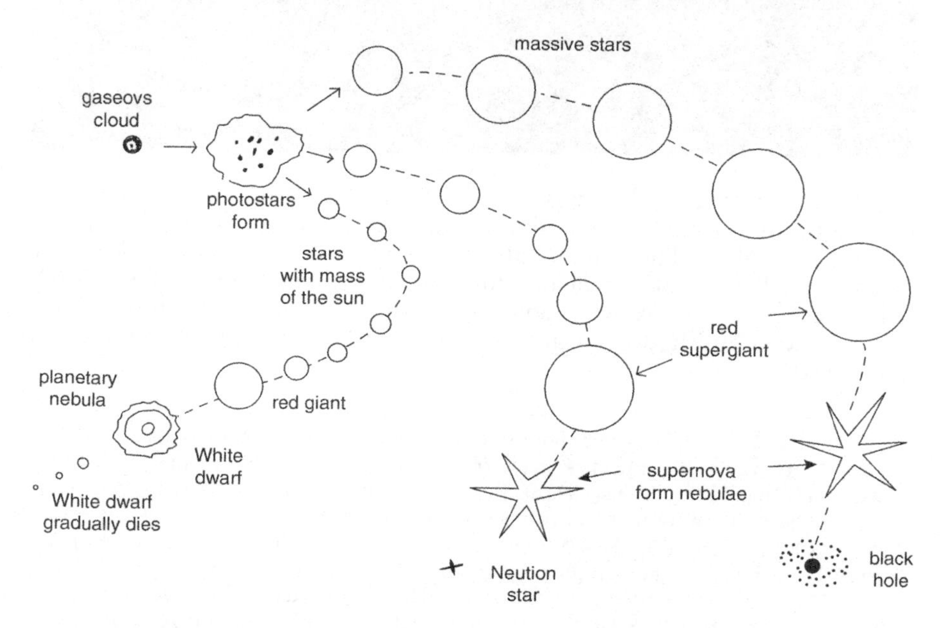

Fig. 10.5 Three pathways for stellar evolution. The path depends on the initial mass of the star.

Fig. 10.6 The ring or planetary nebula (NGC 7293/C63) in the constellation Aquarius formed when a dying star ejected matter that has ionised, causing it to glow. The ring nebula is expanding into space at 25 km/s. It has a central star that is probably forming into a *white* dwarf. The *bluish* colour comes from oxygen ions, while the *pink* and *red* comes from nitrogen ions and hydrogen atoms. The nebula is about 700 light years from Earth.

Sirius B was originally a star about five times the mass of our Sun, before evolving into a white dwarf. Its current size (12,000 km diameter) is about the same as the Earth, but it has as much mass as the Sun packed into its volume. White dwarfs can have a mass of no more than 1.4 times the mass of the Sun.

Sirius B has a spectral class of DA2 and a surface temperature of 25,000 K. It is currently in the cooling down phase of its life. This tiny star is primarily composed of a carbon-oxygen mixture that was generated by helium fusion; however, the outer atmosphere of Sirius B is almost pure hydrogen. Sirius B spins on its axis at an incredible 23 times a minute, generating an enormous magnetic field.

Sirius itself (known as Sirius A) is over 20 times brighter than our Sun and over twice as massive. Sirius A is a main sequence star with spectral class A1 V and surface temperature of 9,940 K. Sirius A itself will one day evolve into a white dwarf star.

Sirius B traces an elliptical orbit around Sirius A, and the two stars take 50 years to complete one orbit around their common centre of gravity. The distance separating Sirius A from its companion varies between 8.1 and 31.5 astronomical units (AU). The age of the Sirius star system has been estimated at around 230 million years. See Fig. 10.7.

Fig. 10.7 This Hubble Space Telescope image shows Sirius A, the brightest star in our night time sky, along with its faint, tiny stellar companion, Sirius B. Astronomers overexposed the image of Sirius A (at *centre*) so that the dim Sirius B (tiny dot at lower *left*) could be seen. The cross-shaped diffraction spikes and concentric rings around Sirius A, and the small ring around Sirius B, are artefacts produced within the telescope's imaging system (Credit: NASA/H.E. Bond and E. Nelan (Space Telescope Science Institute, Baltimore, MD); M. Barstow and M. Burleigh (University of Leicester, U.K.); and J.B. Holberg (University of Arizona)).

The Fate of the Sun

There are several ways in which stars can end their lives, and it depends on how much mass they have to start with. In the final stages of its life, about five billion years from now, our Sun will expand in size to become a red giant. The Earth and other inner planets will be engulfed by the expanding Sun and will cease to exist. The Sun will throw off its outer layers, and the gas will spread outward eventually forming a planetary nebula. What will be left will be a white dwarf star about the size of Earth. This white dwarf will not be able to sustain nuclear fusion – all it will have left is its residual heat, which it will lose over billions of years. It will cool down slowly and eventually become a 'black dwarf'.

Will our Sun ever explode as a supernova? For a white dwarf to become a supernova it requires a companion star in close binary orbit. Our Sun does not have a companion star, so it will not become a supernova. A white dwarf with a companion star pulls some of the ordinary star's gas onto itself, slowly building it up until the white dwarf's mass reaches a critical limit, around 1.4 times the mass of the Sun. At this point, the white dwarf can't handle the increased pressure. It becomes unstable and undergoes a huge nuclear explosion that destroys it (Fig. 10.8).

So for the next several billion years, humanity is safe – in terms of the Sun's existence, at least.

A Magnificent Sun

To the uninstructed eye, the Sun simply appears to be an intolerably bright disc in the sky. To others it is a glorious sight as it rises every morning in the east and sets every night in the west. In any case, it is naturally regarded as the most magnificent of all the heavenly bodies, and the one, which most interests us personally, since all forms of terrestrial life are dependent on it.

The Sun in fact, is a star, one among the thousands of millions that populate the universe. All the stars apart from the Sun appear to us as mere points of light. If the effects of unequal distances are allowed for, the real brightness of the Sun and the stars can be compared, and it is then that we find that our own star occupies a

Fig. 10.8 One of the most famous supernova explosions formed the Crab Nebula in the constellation Taurus. Chinese astronomers saw the explosion in AD 1054. Today the nebula is about 10 light years in diameter and is expanding at about 1,500 km/s. It is located about 6,300 light years from Earth. In the nebula's very centre lies a pulsar: a neutron star as massive as the Sun but with only the size of a small town. The Crab pulsar rotates about 30 times each second (Credit: Hubble Space Telescope 2008 image; NASA/ESA, J. Hester, A. Loll (ASU)).

very modest position in the stellar hierarchy. But it is still magnificent to most people living on Earth.

Hopefully this book will inspire you to continue to study the Sun in all its glory. Clear skies and good, safe observing!

Web Notes

For a list of the largest known stars see: http://en.wikipedia.org/wiki/List_of_largest_known_stars

For information about the Sirius binary star system see Wikipedia: http://imagine.gsfc.nasa.gov/docs/science/know_l2/stars.html

For information about types of supernova try: http://imagine.gsfc.nasa.gov/docs/science/know.../supernovae.html

A list of the latest Supernovae with reference images can be found at: http://www.rochesterastronomy.org/supernova.html

Glossary

Absolute luminosity The amount of energy radiated per unit time, measured in watts.

Active region The magnetised region in, around and above sunspots.

Alfven waves Transverse waves that travel through electrically conducting fluids or gases in which a magnetic field is present, like the Sun's plasma.

Angstrom A unit of length equal to 10^{-10} m, or 0.1 nm, used to measure wavelength.

Annular eclipse A solar eclipse in which a ring (annulus) of solar photosphere remains visible.

Aphelion The farthest distance from the Sun in the elliptical orbit of a planet, comet or asteroid.

Apparent magnitude The visible brightness of a star or planet as seen from Earth.

Asteroid A small rocky and/or metallic object with a small size, often irregular, orbiting the Sun in the Asteroid Belt.

Astronomical unit (AU) The mean distance between the Earth and the Sun, about 150 million kilometres.

Atmosphere A layer of gases surrounding a star, planet or moon, held in place by gravity.

Atom The smallest particle of an element that has the characteristics of the element and is composed of protons, electrons and neutrons.

Aurora Illuminations of the night sky, known popularly as the northern lights or southern lights. Auroras are produced when energetic electrons ionise atoms in the upper atmosphere of Earth.

Axis The imaginary line through the centre of a planet or star around which it rotates.

Baily's beads Beads of light visible around the rim of the Moon during a total solar eclipse. They result from the solar photosphere shining through valleys at the edge of the Moon.

Bandpass (or bandwidth) Measurement between the lower and upper cut-off wavelength of an optical filter.

Black hole An object in the universe whose gravity is so strong that not even light can escape from it.

Calcium H and K lines Spectral lines of singly ionised calcium denoted by Ca II, in the violet part of the spectrum at 3,968 A (H) and 3,934 A (K).

J. Wilkinson, *New Eyes on the Sun*, Astronomers' Universe,
DOI 10.1007/978-3-642-22839-1,

CCD Charge coupled device, a solid state imaging device.

Celestial sphere The imaginary sphere surrounding the Earth, upon which the stars, galaxies, and other objects all appear to lie.

Chromosphere The layer of the solar atmosphere lying between the photosphere and corona.

Comet A small body composed of ice and dust that orbits the Sun on an elliptical path.

Constellation One of 88 officially recognised patterns or groups of stars in the sky.

Continuum That part of the spectrum that has neither absorption nor emission lines but only a smooth wavelength distribution of intensity. The coloured 'rainbow' part of the visible spectrum.

Convection A method of heat energy transfer in a fluid.

Convective zone A layer in the Sun or star where energy is transported outward by means of convection.

Core The innermost region or central region of a planet or star.

Corona The outermost, high temperature region of the Sun's atmosphere above the chromosphere.

Coronagraph An instrument used for observing the solar corona outside a solar eclipse.

Coronal hole Region of the corona that appears dark when observed in the extreme ultraviolet and soft x-ray region of the electromagnetic spectrum.

Coronal mass ejection Abbreviated CME, a transient ejection of plasma and magnetic fields at high speed from the Sun's corona into interplanetary space.

Coronal streamers Wisp-like stream of particles travelling outwards through the Sun's corona. They are visible during a total solar eclipse or in images taken with a coronagraph.

Cosmic rays Nuclear particles or nuclei travelling through space at high velocity and originating outside the solar system.

Declination The angular distance of a celestial body north (+) or south (−) of the celestial equator.

Degree A unit of angular measure, equal to 1/360 of a circle.

Density The amount of mass in an object per unit volume. Usually measured in kg/m^3 or g/cm^3.

Differential rotation The rotation of a gaseous body, such as the Sun, where the rotation rate varies with latitude.

Disc or disk The visible, circular part of the Sun.

Doppler effect Change in wavelength of a spectral line due to the relative motion between the observer and emitter.

Earthquake A sudden vibratory motion of the Earth's surface.

Eclipse The total or partial disappearance of a celestial body in the shadow of another. Examples: solar eclipse, lunar eclipse.

Electromagnetic radiation (spectrum) The name given to a range of radiations, which travel at the speed of light. Includes infrared rays, ultraviolet rays, visible light, x-rays and gamma rays.

Electron A negatively charged sub-atomic particle.

Equator The imaginary line surrounding a celestial body, half way between its two poles.

Etalon An optical filter that operates by the multiple-beam interference of light, reflected and transmitted by a pair of parallel flat reflecting plates. Used in H-alpha telescopes.

Extreme ultraviolet Abbreviated EUV, a portion of the electromagnetic spectrum from approximately 100–1,000 A.

Faculae Faculae are bright, prominent patches of material seen in the Sun's photosphere. They are an indication of a forming sunspot and show up particularly well through white-light solar filters.

Fibrils Are dark protrusions or horizontal wisps of gas similar to spicules but with about twice the duration seen in H-alpha light.

Filament A feature on the solar surface seen in H-alpha light as a dark wavy structure. It is a prominence projected onto the solar disc.

Flare A sudden and violent release of energy and matter within a solar active region.

Fusion A nuclear reaction in which two light nuclei (such as hydrogen) join to produce a heavier nuclei (such as helium) with the release of energy.

Gamma rays Electromagnetic radiation with the highest frequency and shortest wavelength (1 A).

Geomagnetic storms Geomagnetic storms are disturbances in the Earth's magnetic field caused by gusts in the solar wind that blows by Earth.

Granulation A mottled, cellular pattern visible at high resolution over the surface of the photosphere.

Granule The top of a rising column of gas, originating deep within the convection zone of the Sun.

Gravity A universal force of attraction that exists between two masses.

Greenhouse effect The process by which absorption and emission of infrared radiation by gases in the atmosphere warm a planet's lower atmosphere and surface.

H-alpha light The spectral line of hydrogen at wavelength 6,563 A or 656.3 nm.

Helioseismology The study of sound waves that propagate through the Sun's interior.

Heliosphere A vast region formed in interplanetary space by the solar wind and the Sun's magnetic field.

Hertzsprung-Russell (HR) diagram A plot of the absolute magnitude or luminosity of stars versus their surface temperatures or spectral classes.

Hubble Space Telescope An astronomical telescope in orbit around Earth.

Hydrostatic equilibrium A condition of stability inside a star or planet whereby the inward pull of gravity is balanced by the outward force of gas and radiation pressure.

Infrared radiation Form of electromagnetic radiation beyond the visible light region of the spectrum (7,000 A to 1 mm).

Ion An electrically charged atom that has gained or lost one or more electrons, thus making it positive or negative.

Ionisation The process by which a neutral atom or molecule becomes electrically charged.

Kelvin A unit in temperature in which zero Kelvin equals –273.15°C or absolute zero.

Kuiper belt A doughnut shaped ring of space around the Sun beyond Pluto containing many frozen comet bodies.

Lagrangian point The point about 1/100 of the way from Earth to the Sun, where the gravitational pull of the Earth and Sun balance each other in such a way as to give an orbit of exactly one Earth year.

LASCO Acronym for the Large Angle Spectroscopic Coronagraph on the SOHO space probe.

Latitude Number of degrees north or south of the equator measured from the centre of a coordinate system.

Light year The distance light travels in a vacuum in 1 year.

Limb darkening An effect whereby the edge of the Sun's disc appears darker than the central region of the Sun's disc.

Luminosity Absolute brightness of a glowing body such as a star. It is the amount of energy radiated per unit time by the object. Measured in watts where 1 W = 1 J/s.

Magnetic field A region of force surrounding a magnetic object. Generated by moving electric currents.

Magnetic reconnection A process by which magnetic field lines are broken and then rejoined into a new configuration.

Magnetic storm A disturbance in the Earth's magnetic field due to the passage of extra energetic particles from the Sun.

Magnetogram A picture or map of variations in the magnetic field across the Sun's surface.

Magnetosphere The region surrounding a planet or star in which its magnetic field has an effect.

Magnitude A measure of the brightness in the sky of a celestial object.

Main sequence star A star, fusing hydrogen from helium in its core, whose surface temperature and luminosity place it on the main sequence of the Hertzsprung-Russell diagram.

Mantle An inner region of a planet or moon that lies between its outer crust and central core.

Mass The amount of material in a body. Usually measured in grams or kilograms.

Nanometer (nm) A unit used to measure wavelengths of electromagnetic radiation. 1 nm $= 10^{-9}$ m.

Nebula A cloud of interstellar dust and gas found in the universe.

Neutrino A spinning, sub-atomic particle with no electric charge and very little rest mass.

Neutron A sub-atomic particle with no electric charge found in all atomic nuclei except hydrogen-1.

Neutron star A very compact, dense stellar remnant composed almost entirely of neutrons.

Nuclear fusion A process whereby light atomic nuclei (e.g. hydrogen or helium) combine to produce heavier nuclei, with the release of energy. Occurs in stars but not planets.

Oort cloud A spherical region of the solar system beyond the Kuiper belt where most comets are believed to spend most of their time.

Orbit Path of one celestial body when revolving around another.

Ozone layer The lower stratosphere, where most of the ozone in the air exists.

Penumbra The lighter, greyish outer region surrounding the umbra of a sunspot.

Perihelion The closest distance to the Sun in the elliptical orbit of a comet, asteroid or planet.

Period The time taken for a planet to orbit the Sun or a moon to orbit a planet.

Photosphere The lowest layer of the Sun's atmosphere. Contains sunspots, granulation and faculae.

Plage A bright region that often surrounds sunspots. It is associated with the magnetic field of sunspots.

Planet A celestial body that is in orbit around the Sun, has sufficient mass for its self-gravity to overcome rigid body forces so that it assumes a hydrostatic equilibrium (nearly round) shape, and has cleared its neighbourhood around its orbit.

Plasma An ionised gas consisting of electrons and ions.

Pore A small, short-lived dark area in the photosphere out of which a sunspot may develop.

Positron A positively charged electron.

Prominence A flame-like structure seen around the limb of the Sun and extending into the corona. When viewed against the surface, it appears darker and is called a filament. It is made of high density gas.

Proton A sub-atomic particle with a positive charge. Located in the nucleus of the atom.

Protostar A star in its early stages of formation.

Protosun The part of the solar nebula that eventually developed into the Sun.

Pulsar A pulsating radio source associated with a rapidly rotating neutron star.

Radiative zone An inner layer of the Sun in which energy transfer occurs through radiative properties.

Radio blackouts Radio blackouts are disturbances in the ionosphere caused by x-ray emissions from the Sun.

Radio telescope A telescope or antennae designed to detect and receive radio waves from outer space.

Radio waves Electromagnetic waves of long wavelength – from 0.001 m to 30 m.

Revolution The motion in which a body moves around another body, such as the Moon around the Earth or the planets around the Sun.

Rotation The spin of a planet, moon, satellite or star on its axis.

Satellite Any small object (artificial or natural) orbiting a larger one, such as a body orbiting a planet.

SDO Acronym for the Solar Dynamics Observatory.

SOHO Acronym for the Solar and Heliospheric Observatory.

Solar cycle An 11 year rise and fall in solar activity, e.g. Sunspot cycle.

Solar eclipse A blockage of light from the Sun when the Moon passes precisely between the Sun and Earth.

Solar flare A sudden release of matter and energy from an active region on the Sun.

Solar irradiance Solar irradiance is the amount of radiant energy emitted by the Sun over all wavelengths as measured outside Earth's atmosphere.

Solar maximum The peak of solar activity during the 11 year solar cycle.

Solar minimum The time during which the Sun experiences little activity during the 11 year solar cycle. Marks the beginning or end of sunspot cycle.

Solar nebula The cloud of gas and dust from which the Sun and Solar System formed.

Solar radiation storms Are storms with elevated levels of radiation that occur when the numbers of energetic particles increase.

Solar system The Sun, planets and their satellites, asteroids, comets, and related objects that orbit the Sun.

Solar wind A stream of charged particles (mainly electrons and protons) or ions emitted by the Sun and thrown into space.

Space probe A spacecraft or artificial satellite used to explore other bodies (such as the Sun, planets or Moon) in the Solar System. Such a craft contains instruments to record and send back data to scientists on Earth.

Space weather The changing conditions in near-Earth space within the magnetosphere and ionosphere, but it is also studied in interplanetary (and occasionally interstellar) space.

Spectroscopy The analysis of light from a planet or star to determine its composition and condition.

Spicule A small jet of gas seen at the edge of the Sun in the chromosphere.

Star A giant sphere of plasma that releases its own heat and light via nuclear fusion reactions.

Stellar parallax The apparent shift in a nearby star's position on the celestial sphere resulting from the Earth's orbit around the Sun.

Sunspot A dark, cooler region on the Sun's photosphere with strong magnetic fields.

Supergranulation cells Large convective cells covering the Sun's photosphere.

Supernova A stellar outburst during which a star suddenly increases its brightness roughly a million fold.

Tachocline A region inside the Sun between the radiative zone and the convective zone.

Transit The passage of one celestial body in front of another celestial body.

Transition region A thin region of the Sun's atmosphere between the chromosphere and the corona.

Troposphere The lowest level of the Earth's atmosphere.

Ultraviolet radiation Electromagnetic radiation with a shorter wavelength than visible light (100–3,500 A). Extreme ultraviolet radiation (EUV) lies in the shorter wavelength part of the range.

Umbra The dark inner core of a sunspot visible in white light.

Van Allen radiation belts Two flattened, doughnut shaped regions around the Earth where many charged particles are trapped by the Earth's magnetic field.

Visible light Electromagnetic radiation with wavelength 4,000–7,000 A. See white light.

Wavelength The distance between successive crests or troughs in a wave. Wavelength is inversely proportional to frequency.

White light The visible portion of sunlight that includes all of its colours. See Visible light.

Weight The force of gravity acting on an object.

X-rays Electromagnetic waves with a short wavelength between ultraviolet and gamma rays, range 1–100 A.

Zeeman effect A splitting or broadening of spectral lines in the presence of a magnetic field.

About the Author

John Wilkinson is a science educator with over 30 years experience in teaching science, physics and chemistry in secondary colleges and universities in Australia. He is author of over 100 science textbooks. He completed his Masters degree and PhD in science education at La Trobe University, Australia. Throughout his life he has been a keen amateur astronomer and operates his own observatory from his backyard. His main astronomical interests include the Moon, Sun and Solar System objects.

John is also author of "Probing the new solar system" and "The Moon in close-up". His web site is http://astroscimac.com.

J. Wilkinson, *New Eyes on the Sun*, Astronomers' Universe,
DOI 10.1007/978-3-642-22839-1, © Springer-Verlag Berlin Heidelberg 2012

Index

J. Wilkinson, *New Eyes on the Sun*, Astronomers' Universe,
DOI 10.1007/978-3-642-22839-1, © Springer-Verlag Berlin Heidelberg 2012

Made in United States
North Haven, CT
20 September 2024